C0-AYI-685

Climate Change and Economic Development

Other books by Sardar M.N. Islam

Sardar M.N. Islam and Christine Mak (*authors*)
NORMATIVE HEALTH ECONOMICS
A New Pragmatic Approach to Cost Benefit Analysis, Mathematical Models and Applications

M. Rusydi and Sardar M.N. Islam (*authors*)
QUANTITATIVE EXCHANGE RATE ECONOMICS IN DEVELOPING COUNTRIES
A New Pragmatic Decision Making Approach

Climate Change and Economic Development

SEA Regional Modelling and Analysis

Jamie Sanderson and Sardar M.N. Islam

First published 2007 by
PALGRAVE MACMILLAN
Houndmills, Basingstoke, Hampshire RG21 6XS and
175 Fifth Avenue, New York, N.Y. 10010
Companies and representatives throughout the world

PALGRAVE MACMILLAN is the global academic imprint of the Palgrave Macmillan division of St. Martin's Press, LLC and of Palgrave Macmillan Ltd. Macmillan® is a registered trademark in the United States, United Kingdom and other countries. Palgrave is a registered trademark in the European Union and other countries.

ISBN 13: 978–0–230–54279–2 hardback
ISBN 10: 0–230–54279–4 hardback

This book is printed on paper suitable for recycling and made from fully managed and sustained forest sources. Logging, pulping and manufacturing processes are expected to conform to the environmental regulations of the country of origin.

A catalogue record for this book is available from the British Library.

A catalogue record for this book is available from the Library of Congress.

10 9 8 7 6 5 4 3 2 1
16 15 14 13 12 11 10 09 08 07

Printed and bound in Great Britain by
Antony Rowe Ltd, Chippenham and Eastbourne

For Grace and Max

Contents

List of Tables

List of Figures

Acknowledgement and Sources of Some Materials

The authors thank the publishers of the following articles and chapters for allowing these materials to be published in this book:

1. Sanderson, J. and S.M.N. Islam (2001), Economic Development and Climate Change in South East Asia: The SEADICE model and its Forecasts for Growth Prospects and Policy Strategies, *International Journal of Global Environmental Issues,* Vol. 3, No. 2, 2003, Inderscience.
2. Sanderson, J. and S.M.N. Islam (2006), Scenarios for Climate Change in South East Asia: Implications for Economic Development and Policy Options, *International Journal of Environmental Creation,* Vol. 5, No. 2.
3. Sanderson, J. and S.M.N. Islam (2000), Economic Growth and Climate Change in Asia: Costs, Issues and Policy Options, with Jamie Sanderson, *Natural Resources Forum*, United Nations, New York, Elsevier Science/Blackwell Publishing.

The authors also thank the following publishers for allowing the following materials to be published in this book:

1. Earthscan Publishers for Table 3.2 Types of Impacts Resulting from Climate Change.
2. Kluwer Law International for Table 2.4 Environmental Conventions to which South East Asian Countries are Party.

Summary and Preface

The economic impacts of climate change have the potential to be unevenly distributed around the globe, as stressed in the recent Stern Report (Stern, Peters, Bakhshi, et al. 2006). In particular, the developing regions of the world will likely be the most vulnerable to the impacts of climate change. In this book the region of South East Asia is the focus with respect to the economics of climate change, an area of economics that is subject to great uncertainties with respect to data, model specification and results. South East Asia is a rapidly developing region both economically and socially, and in an environmental sense the countries of the region encompass similar ecosystems and climatic conditions. Both of these factors make it an interesting region for study. This book examines the region's vulnerability to the impacts of climate change, forecasts the environmental and economic outcomes for the region arising from its vulnerability and also the opportunities these factors provide for policy actions towards alleviating climate change vulnerability, particularly through adaptation.

From a collection of regional sectoral data on the various potential impacts of climate change an aggregate impact estimate for South East Asia is made in this book which indicates that economic output will be reduced by 5.3% for climate change conditions where atmospheric concentrations of carbon dioxide are double pre-industrial levels. This estimate and other region specific data for South East Asia are used to implement the South East Asia Dynamic Integrated Climate and Economy (SEADICE) or the South East Asian DICE model, a dynamic optimisation integrated model of climate and economics, which is based on the DICE model (Nordhaus and Boyer 2000). This model provides dynamic optimal forecasts of environmental and economic variables so that the full impacts of climate change can be observed. It also provides the basis for a novel experiment where through the application of endogenous growth theory, the SEADICE model is modified to incorporate endogenous technical growth.

After a review of the theoretical literature on climate change adaptation and based on arguments made in the book, endogenous technical growth is assumed to represent autonomous adaptation to climate change. The experiment is successful and while much like the rest of this literature the results are illustrative they do indicate that when various levels of autonomous adaptation are given in the SEADICE model they produce substantial changes in economic output. Therefore, the consequences of autonomous adaptation may be important and should be the subject of further investigation.

Using the modelling results, arguments and conclusions from throughout the book, climate change policy recommendations are made. As South East Asia is a region of developing countries, the following two courses of action are recommended for mitigation policy over the next decade. (1) The pursuit of Clean Development Mechanism projects with Annex I countries that will deliver foreign direct investment and technology transfer benefits and (2) focus on no regrets mitigation policy options through demand side management techniques, concentrated in the energy and forestry sectors of the economy. For adaptation policy options it is recommended that adaptation policies be identified on a regional basis through an institution such as ASEAN, which can utilise the pooled expertise throughout the region. The policy recommendations for both mitigation and adaptation are not controversial, however in this instance they have been supported by research based on two aspects of the literature; climate change impacts on a developing region and climate change adaptation. Both aspects at this stage have not been attempted using a dynamic optimal control integrated model for the region of South East Asia to the authors' knowledge.

The first author would like to thank some of the people who are related to him, both academically and socially: his family including his wife Julie Atkinson, children Grace and Max, parents Marie and Russell Sanderson and Kylie Sanderson. He also thanks staff at the Centre for Strategic Economic Studies including Professor Peter Sheehan and Margarita Kumnick. Both the authors thank Dr. Ruchi Gupta and Kashif Rashid for providing editorial assistance in preparing the final version of the paper. The authors appreciate the support and encouragement provided by Amanda Hamilton and Alec Dubber of Palgrave Macmillan for publishing this book.

Foreword

Climate change remains as one of the challenges in the field of environmental and development economics. Developing countries/regions like the South East Asian region are at greater risk from any climate change as it conflicts with the policy of sustainable development which is crucial for any country in this region. To enable economic development to proceed in a sustainable manner it is important to address the issues of climate change and its impact on economic development. This book focuses on the region of South East Asia with respect to the economics of climate change and economic development – an area of economics that is subject to great uncertainties with respect to data, model specification and results. It examines the impacts of climate change for the region and forecasts the long run environmental and economic outcomes. Adaptation related policy changes as an option to alleviate climate change problems tend to get ignored in any policy change initiative.

This book has covered several important areas of climate change economics including macroeconomic impact estimates, climate change economic modelling and the theoretical and practical aspects of the economics of adaptation to climate change with a focus on South East Asia.

It is an important book addressing the most crucial issue facing the human community at the moment. It presents some significant research contributions to this area of literature and policy dialogue through impact estimates, regional analysis of theoretical and practical aspects of the economics of adaptation to climate change, regional modelling, endogenous growth analysis, and policy prescriptions.

This book should be of great interest to development and environmental policy makers, NGOs, academics, practitioners, researchers, and students of climate change and economics.

Dr. Liana BRATASIDA
Assistant Minister for Global Environment Affairs
and International Cooperation,
Ministry of Environment,
Government of Indonesia.
and
National Focal Point – Indonesia,
Asia-Pacific Network for Global Change Research (APN),
Kobe, Japan.

March 2007

List of Abbreviations

$US	United States Dollars
ADB	Asian Development Bank
ADICE	Australian Dynamic Integrated model of the Climate and Economy
ALGAS	Asia Least-cost Greenhouse gas Abatement Strategy
APEC	Asia-Pacific Economic Cooperation
APF	Adaptation Policy Framework
APN	Asia Pacific Network for Global Change Research
ASEAN	Association of South East Asian Nations
ASEP	ASEAN Environment Program
ASOEN	ASEAN Senior Officials of the Environment
CAS	Complex Adaptive System
CDM	Clean Development Mechanism
CER	Certified Emission Reduction
CO_2	Carbon Dioxide
COP	Conference of the Parties
CVI	Coastal Vulnerability Index
DARLAM	Division of Atmospheric Research Limited Area Model
DICE	Dynamic Integrated Climate and Economy Model
DSM	Demand Side Management
EKC	Environmental Kuznets Curve
FAR	First Assessment Report
FARM	Future Agricultural Resources Model
FCCC	Framework Convention on Climate Change
GCM	Global Circulation Model
GDP	Gross Domestic Product
GEF	Global Environment Facility
GHG	Greenhouse Gases
GIM	Global Impact Model
GNP	Gross National Product
Gt/C	Gigaton of Carbon
Ha	Hectares
HDRUE	Higher Decline Rate of Uncontrolled Emissions
HMS	Hydrometeorological Service

IAM	Integrated Assessment Model
IC-SEA	Impacts Centre for South East Asia
ICSU	International Council of Scientific Union
IPCC	Intergovernmental Panel on Climate Change
JI	Joint Implementation
km	Kilometres
LMI	Lower Middle Income
m	Metre
MIT	Massachusetts Institute of Technology
Mt	Megatons
NGO	Non Government Organisation
NPV	Net Present Value
OE	Operational Entity
OECD	Organization for Economic Cooperation and Development
R&D	Research and Development
ROW	Rest Of the World
SAR	Second Assessment Report
SDR	Social Discount Rate
SEA	South East Asia
SEADICE	South East Asia Dynamic Integrated Climate and Economy model
SLR	Sea Level Rise
START	SysTem for Analysis, Research and Training
TAR	Third Assessment Report
TFP	Total Factor Productivity
Tg	A Million Tons
UN	United Nations
UNDP	United Nations Development Programme
UNEP	United Nations Environmental Programme
UNFCCC	UN Framework Convention on Climate Change
USCSP	United States Country Studies Program
WGI	Working Group I
WGII	Working Group II
WGIII	Working Group III
WMO	World Meteorological Organization
WTP	Willingness to Pay
YLL	Years of Life Lost

1
Introduction: Issues and Developments in Climate Change

1.1 Introduction

Since the peak of public awareness in the early 1990s the global environmental problem known as climate change[1] has been developing and transforming rapidly in recent years, as evidenced in the recent Stern Report (Stern, Peters, Bakhshi, et al. 2006). One of the most important factors has been the transformation of adaptation from an issue that early on was either ignored or given as an afterthought in the overall climate change debate, whereas now it is a much more significant theoretical and policy issue. This book examines the theoretical and practical aspects of adaptation to climate change from an economic perspective. The basis for this analysis is the implementation of a dynamic optimisation integrated model that will be used to explore and expand upon the issues and arguments that will be raised in this book. The model implemented here will be based upon the Dynamic Integrated Climate and Economy model (DICE) of Nordhaus (see Nordhaus and Boyer 2000) and its geographical scope will be the region of South East Asia (SEA). This type of modelling is still developing and is burdened with high levels of uncertainty. If the precautionary principle is followed,[2] although this type of climate change research involves highly uncertain outcomes, climate change is a serious global concern and the small potential for catastrophic outcomes justifies the general research done in the climate change area and certainly the research undertaken in this book. In order to provide a context upon which later arguments are based the following section reviews the historical developments in the broad climate change literature.

1.2 A brief review of the science and history of climate change

The science and history of the greenhouse effect has previously been described in many publications (Nordhaus 1994a; Fankhauser 1995b; Janssen 1996). Therefore, for the purposes of this document the review of the science and historical discovery of climate change will be kept quite brief. This review serves as background information for this book.

1.2.1 The discovery of the greenhouse

The natural greenhouse effect was first described by French physicist Jean Baptiste Fourier in 1824. Fourier hypothesised that the Earth's atmosphere acted similarly to a greenhouse, with certain gases in the atmosphere trapping some of the heat from the sun's radiation rather than allowing all of the radiation to bounce back into space (Fourier 1824). Tyndall (1863) was another who examined this issue. The greenhouse analogy is used because the greenhouse and the atmosphere both have a clear surface that lets heat in yet which is relatively opaque with respect to the infrared radiation reflected from the surface of the Earth, so that a certain amount of heat is retained. As a result of this 'greenhouse effect' the Earth's atmosphere is many degrees warmer than it otherwise would be without the existence of the greenhouse gases (GHG) and is therefore vital for the current abundance of life. It was Svante Arrhenius in 1896 who first speculated that the burning of coal by humans could be a contributing factor towards increased carbon dioxide (CO_2) concentrations in the atmosphere and a subsequent warming of the climate.[3] Arrhenius predicted that surface temperatures would rise 9°C as a result of a threefold increase in pre-industrial CO_2 concentrations, this prediction is unerringly similar to many of today's estimates. These two theories did not create an immediate concern for their possible consequences[4] and were really only rediscovered after further developments many decades later, instead they represent the eclectic first hint of climate change science.

1.2.2 The international response to climate change

While the recognition of the 'greenhouse' phenomenon has spanned almost two centuries, the recognition of it as a potential international problem spans just the last several decades. In 1957, Revelle and Suess disputed the widely held view that the CO_2 balance between the oceans and the atmosphere was stable and that the ocean absorbed the CO_2 emissions of industrialised society (Revelle and Seuss 1957). This led to Revelle and Seuss suggesting that monitoring of the level of CO_2 in the atmosphere should be conducted. Consequently, the monitoring station on Muana Loa, Hawaii was commissioned where it was discovered that GHG concentrations were in fact rising. The evidence for the potential for anthropogenic climate change was beginning to mount.

In 1965 a chapter in the United States President's Science Advisory Committee report was devoted to atmospheric CO_2 (PSAC 1965). This was significant as it was the first government sanctioned document to explicitly address the issue. Other studies started to appear, particularly around the late 1960s to early 1970s, such as the 1970 Massachusetts Institute of Technology (MIT) report 'Man's Impact on the Global Environment' where substantial portions of the document were devoted to the serious measurement of possible human induced climate change (SCEP 1970). A 1972 conference in Geneva, the First World Climate Conference, expressed the desire for increased efforts towards more research on the topic of climate change. However, the real basis for serious international involvement in climate change stems from an international conference in 1985 in Austria organised by the United Nations Environmental Programme (UNEP), the World Meteorological Organization (WMO) and the International Council of Scientific Union (ICSU). For the first time a substantial international consensus involving high level international organisations and amongst scientists was reached, on the conclusion that in the first half of the 21st century a global mean temperature rise of a magnitude greater than at any previous time in human civilisation could occur. The seriousness of the problem was now apparent to the wider international scientific community.

In 1988 the Intergovernmental Panel on Climate Change (IPCC) was formed at a meeting of 35 countries initiated by UNEP and the WMO. The primary purpose of the IPCC was the provision of authoritative assessments to governments of the state of knowledge concerning climate change. Three working groups were established under the mission of the IPCC, namely:

1. The science of climate change (Working Group I).
2. Impacts, adaptation and mitigation options (Working Group II).
3. Economic and social dimensions (Working Group III).

The IPCC is precluded from making policy recommendation to governments, as its primary purpose is to form the knowledge upon which others can base their own policy decisions. The First Assessment Report (FAR) was produced in 1990 with results from all three working groups. Working Group I (WGI) concluded that rising concentrations of GHG in the atmosphere were caused by human activities which might have subsequent effects on future climate scenarios, but significant uncertainty existed. The central forecast for the increase in the global mean surface temperature of 0.3°C (±0.15°C) per decade is faster than at any time in human civilisation. The results from Working Group II (WGII) were subjected to widespread uncertainty and disagreement within the group. The main point of contention came from the uncertainty of local climate change effects in the future. Broadly, the conclusions were that sea level rise (SLR) and rainfall distribution would be major effects of climate change and that the impacts on sectors such as agriculture, coastlines, forest and wetlands could be significant. The results from Working Group III (WGIII) were the subject of intense political negotiations as its charter was a review of the possible responses to climate change. The only substantial outcome was a recommendation to start negotiations, for a global agreement on a climate change response.

The next major event was the 1992 Earth Summit in Rio De Janeiro, which was the culmination of negotiations for over one year to develop a UN Framework Convention on Climate Change (UNFCCC). When the agreement was signed, the final aim was for the stabilisation of GHG concentrations that would prevent harmful anthropogenic interference with the climate system within a time frame that would allow ecosystems to adapt naturally. '[To achieve] stabilisation of GHG concentrations in the atmosphere at a level that would prevent dangerous anthropogenic interference with the climate system. Such a level should be achieved within a time-frame sufficient to allow ecosystems to adapt naturally to climate change, to ensure that food production is not threatened and to enable economic development to proceed in a sustainable manner' (Framework Convention on Climate Change, Article 2). This agreement was not binding and the only specific recommendation was for Annex I countries to reduce their emissions of GHG to 1990 levels by the year 2000.[5] The convention finally came into force in March 1994 with over 160 countries as signatories.

In 1995 the Second Assessment Report (SAR) was produced by the IPCC. Working Group I measured the levels of GHG in the atmosphere and made estimates of their growth in recent decades. More knowledge was gained about the chemistry of the atmosphere such as the role aerosols play in the cooling of the atmosphere (Brack and Grubb 1996). The main conclusion is significant and has been widely quoted throughout the relevant literature, 'the balance of evidence suggests that there is a discernible human influence on global climate' (IPCC 1996a). For the first time it was acknowledged by a wide cross section of international scientists that climate change is a real problem and that human systems were the primary cause.

Working Group II came to the following conclusions: The composition and distribution of ecosystems will shift as a result of climate change. Any changes in the hydrological cycle would be of concern to marginally arid regions of the globe. While aggregate agricultural production is predicted to be maintained, the composition of production is expected to change, which may place marginal agricultural lands in developing nations in potential difficulties. While human infrastructure is less vulnerable to climate change, it is still vulnerable to any change in the variability of climate, such as an increase in storm or cyclone activity (Burton 1997, Dixon 1999). It was also concluded that climate change may have adverse effects on human mortality as a result of heat stress and an increase in the frequency of extreme events and the potential for increases in the transmission of vector-infections such as malaria. Predictions of energy efficiency gains of 10–30% above present levels at little or no cost were estimated for many parts of the world.

The results from WGIII proved to be the most contentious. These include the conclusion that significant opportunities exist in most countries to implement no regrets[6] measures to reduce GHG emissions at no net cost. It was also concluded that policy actions beyond no regrets were justified on the grounds of pursuing the precautionary principle, the existence of risk aversion and the risk of aggregate damage from climate change. It was suggested that sensible governments would consider dealing with climate change through a mixture of mitigation, adaptation and knowledge growth. The working group was firmly on the fence over the issue of the appropriate discount rate to apply to climate change studies. It concluded that a prescriptive approach would yield discount rates in the range 0.5–3% p.a. and a descriptive approach would lead to discount rates of 3–6% p.a. Neither approach was recommended by the working group. The results of the cost estimates of reducing GHG emissions were found to vary widely, depending on the methodology adopted.

Within a few decades climate change had developed from an obscure scientific problem to a major international concern. Concern, which led to the Kyoto Protocol.

1.2.3 The outcome from Kyoto

The Kyoto Protocol is the most significant step to date in one of the most extensive and important international agreements in history. The most significant meeting on climate change thus far; the Third Conference Of the Parties (COP3) was conducted from the 1st to the 10th of December 1997 at Kyoto in Japan. The Kyoto conference was attended by representatives of 160 countries. The objective of COP3 was to obtain an agreement on legally binding emission targets for the Annex I countries. Throughout the conference debate was very rigorous and animated, and at various stages it appeared as though a consensus may not be reached. Following the precautionary principle, prior to the conclusion of COP3 the Annex I countries agreed to reduce their GHG emissions by different amounts that produced an aggregate reduction to 5.2% below 1990 levels, by the year range 2008–12.[7] Table 1.1 below illustrates the variety of emission targets set for the Annex I nations.

Preliminary approval was reached on developing mechanisms to allow emission credits for Annex I nations for the establishment of an emissions trading regime between the Annex I nations. The convention obliges all parties to prepare national inventories of GHG emissions, to prepare national climate change policy pro-

Table 1.1 Kyoto Protocol Emission Targets of Annex I Countries

Party	Emission Target as a Percentage of 1990 Levels
Australia	108
Austria	92
Belgium	92
Bulgaria	92
Canada	94
Croatia	95
Czech Republic	92
Denmark	92
Estonia	92
European Community	92
Finland	92
France	92
Germany	92
Greece	92
Hungary	94
Iceland	110
Latvia	92
Liechtenstein	92
Lithuania	92
Luxembourg	92
Monaco	92
Netherlands	92
New Zealand	100
Norway	101
Poland	94
Portugal	92
Romania	92
Russian Federation	100
Slovakia	92
Slovenia	92
Spain	92
Sweden	92
Switzerland	92
Ukraine	100
United Kingdom	92
United States of America	93

grams, cooperate in research and monitoring and to promote awareness of the issue. The preamble to the Protocol acknowledges the need for developed countries to be the first to take action on climate change.

However, ratification was under threat when in March of 2001 the United States President George W. Bush declared that the United States was abandoning the Kyoto Protocol. The primary reason given for this was that the United States perceived the fact that developing nations are not part of the Protocol to be unfair to the United States. Their major concern is with countries such as China and India that have much smaller per capita emissions but overall emission levels that rival the United States. This was a severe blow to the protocol, which needs the ratification of 55 countries representing over 55% of global emissions to become legally enforceable. The United States represents around 25% of global emissions, therefore the vast majority of other Annex I countries must ratify the treaty for the 55% target to be made possible. As of July 2002 the United States is still refusing to ratify the treaty. While this has caused considerable problems, the protocol is now very close to being implemented. This is due in large part to the successful outcomes at the COP7 meeting at Marrakesh in October of 2001.

The aim of the Marrakesh accord (IPCC 2001d) was to finalise the underlying legal texts for the Bonn Agreement (COP6) and set in place the accounting system for the Kyoto Protocol. This task was completed successfully to finally remove the main barriers to ratification and produce an operational Kyoto Protocol after five years of intense negotiations. The key outcomes were:

1. Eligibility requirements were successfully negotiated for the participation of Annex I countries in the flexibility mechanisms. The requirements for a country to participate are: a) they must be a Party to the Protocol; b) have satisfactorily established its assigned amount; c) have in place its national system for estimating emissions and removals; d) have in place its national registry; e) have submitted its most recent required inventory; and f) submit the 'supplementary information' required to show that it is in compliance with its emissions commitments.
2. The creation of an international accounting system to keep account of all the carbon credits bought and sold and to calculate whether a country has met its target at the end of the commitment period.
3. Penalties were negotiated for the possibility of non-compliance where a country fails to meet its emissions reduction targets. The

following consequences apply: (1) For every ton of emissions by which a Party exceeds its target, 1.3 tons will be deducted from its assigned amount for the subsequent commitment period; (2) the Party will prepare a detailed plan explaining how it will meet its reduced target for the subsequent commitment period; and (3) the Party will not be able to use Article 17 emissions trading to sell parts of its emissions allocation.

As of July 2001, 76 countries had ratified, accepted or approved the Kyoto Protocol. Therefore, over 55 countries have ratified, thus satisfying the first criteria for implementation. However, those that have ratified represent only 36% of 1990 emissions. In order to reach the target of 55% both the Russian Federation (17.4% of 1990 emissions) and Canada (3.3%), are required to ratify the treaty. The secretariat is aiming for the convention to be fully ratified by the end of 2002.

In March 2001 the Third Assessment Report (TAR) of the IPCC was released (IPCC 2001a; 2001b; 2001c). The TAR presents the latest in scientific knowledge on climate change and is the current benchmark. From WGI some of the major conclusions were: (i) global average surface temperature increased by 0.6°C over the 20th century; (ii) globally, 1998 was the warmest year and the 1990s the warmest decade since instrumental records began in 1861 and (iii) there is new and stronger evidence that most of the observable warming in the last 50 years is attributable to human activities. From WGII some of the main conclusions were: (i) many physical and biological systems have already been affected by recent regional climate change; (ii) adaptation is an essential strategy at all scales to complement mitigation efforts and (iii) those regions with the least resources also have the least capacity to adapt to climate change and therefore are the most vulnerable. For WGIII some of the main conclusions were: (i) substantial technical progress has been made since the SAR relevant to GHG reduction and has been faster than anticipated; (ii) conservation and sequestration may allow time for other mitigation options to be developed as significant carbon mitigation potential currently exists for forests, agricultural lands and other terrestrial ecosystems and (iii) countries and regions will have to choose their own path to a low emission future as no single optimal path exists. Overall, the IPCC claimed that significant progress was made in the TAR towards further understanding climate change and the possible human response to it.

It is obvious from the previous discussion of the scientific history and record of international cooperation on climate change that this has been an interesting and monumental process that still has a long road ahead. While this section served as a brief review of the science and history of climate change, this book is concerned with particular aspects of the economics of climate change.

1.3 The economics of climate change

Although to a certain extent the climate change debate was dominated early on by science and politics, economics has been an increasingly important discipline during the development of international efforts to understand and cope with climate change. This book will focus on the branch of climate change economics that deals specifically with the estimation of climate change damage, in particular those that deal with concepts such as adaptation. Other branches such as those devoted to the costs of carbon mitigation will not be referenced in this book. This section briefly reviews how economics is associated with climate change and what it can contribute towards climate change policy solutions.

Economics is very closely associated with climate change and has contributed in both positive and negative ways. Historically, economic growth and development has been the driving force behind the increased release of GHG, which has resulted in a largely negative influence on climate change with respect to emissions. The industrial revolution and its reliance on coal based energy as the driving force behind production is mostly to blame for the build up of GHG in the atmosphere. However, for the present and future, economics should contribute in a more positive manner towards possible solutions to many aspects of the climate change problem. Although up to this point it has been the natural science community that has invested the most resources into examining the nature of climate change, economics will provide many of the potential practical solutions to the problem that are likely to be implemented by high level policy makers.

Economics enables the comparison of costs and benefits of certain aspects of climate change to facilitate the implementation of realistic policy options. Real solutions and information can be provided to policy makers from the application of economics to the climate change problem. Unfortunately economics can also be used effectively for division among negotiating parties. Economic models of climate change can be used to represent and support both sides of the debate depending on their focus. It was not always the case that economics dominated international negotiations, in 1994 Nordhaus stated, 'To date, the calls to arms and treaty negotiations have progressed more or less independently of economic studies of the costs and benefits of measures to slow greenhouse warming' (1994a, p. 4). This has changed in the time since, with many studies completed on the costs of various mitigation policy options, a number of which have been used in international negotiations.

The climate has tangible effects upon an economic system. Although the actual process of industrialisation does tend to insulate the participants over time as they become more isolated from the natural world, climate can never be ignored as an important factor in some sectors of the economy. The IPCC (2001b) identifies some of the economic sectors vulnerable to climate change as water resources, agriculture, transport, forestry, coastal zones, energy, human health, tourism, insurance and other financial services. Of course when the climate effects are severe, such as the case of a drought, the interlinked nature of an economy means that even non-climate sensitive sectors can be affected through secondary and further effects, e.g. the adverse effect on the Australian economy during the drought years of 1982–83.

The problem is to represent climate change as an economic problem where the production and consumption of goods and services today must be optimised so as to minimise the future impacts of climate change. It is in some sense a classic economic problem of allocating scarce resources over time given knowledge of preventable economic damage in the future. In this case however, the time frame is intergenerational, the scale is global and the future economic impact is still highly uncertain. The problems these characteristics present are discussed in the next section.

1.3.1 Problems presented by climate change for economics

Part of the problem presented by global warming is that the atmosphere is a public good and the impact of increased GHG concentrations in the atmosphere might not impact sufficiently upon market mechanisms. As GHG concentrations increase, the externalities they produce are not captured fully by market mechanisms. This is a classic problem in environmental economics (Kneese 1977). However, due to factors such as the reliance on the natural sciences to provide solid physical relationships that can be monetised to obtain confident predictions of climate change impacts the problem is somewhat more complicated. The externalities produced by increased GHG concentrations manifest themselves across time (over generations) and also across space (over regions). Any attempts at an international response have the potential to be undermined by many factors including such things as freeriding behaviour and equity issues. Despite the problems that are present, both theoretical and practical, significant progress is being made in many areas of climate change economics research towards effective solutions to the problems highlighted here (Nordhaus and Boyer 2000; IPCC 2001b; Page 2001). In this book a particular problem presented by climate change will be examined. This problem relates to the way the economy reacts to the impacts of climate change and what policy options are available, given those impacts. The policy option covered in most detail here is that of adaptation to climate change. The definition and measurement of this factor presents a considerable challenge for climate change economics and will be the subject of much of this document.

1.4 Climate change policy options: mitigation and adaptation

One of the most significant aspects of the economics of climate change is the range of policy options that are available to ameliorate the problem. This section serves as an introduction to the concepts of mitigation and in particular adaptation that will be further developed in later chapters.

1.4.1 Mitigation and adaptation: definitions and contrasts

The two major policy options for climate change in broad terms are mitigation and adaptation. By a simple definition, mitigation is the act of reducing GHG emissions with the goal of slowing or preventing climate change, whereas adaptation is the act of reducing vulnerability to the effects of climate change. Mitigation actions are evaluated in terms of cost-effectiveness whereas adaptation measures must be evaluated relative to the benefits they create. The effectiveness of mitigation is measured by a single factor (level of GHG emissions), whereas adaptation effectiveness cannot currently be represented by a single measure. The very simple definitions supplied here demonstrate that whereas mitigation reduces the causes of climate change, adaptation is a reduction in the effect. The clear distinction is between cause and effect. Any action that changes the causes of climate change are mitigation related and any actions that change the effect of climate change are adaptation related.

The first contrast between the two policy issues is the amount of research focus there has been on them. Throughout the development of climate change as an international issue, mitigation has demanded by far the most attention as a policy option. It is currently implemented according to the precautionary principle where uncertainties in potential benefits and costs are ignored since the small possibility exists that climate changes might be catastrophic. Many more economics studies have been done on mitigation and consequently a large range of economic solutions have been raised and are now real policy options at the international level (Kaya et al. 1993).

Another significant contrast exists between mitigation and adaptation (in particular autonomous adaptation, defined and discussed in Chapter 5) in terms of the incentives involved for economic agents. For mitigation to occur, collective action at the national and regional level is needed because the benefits of mitigation are mostly external to individual economic agents and therefore there is no incentive for individuals to undertake mitigation actions. In contrast autonomous adaptation is expected to be realised in an economic sense as the benefits would be internal for individual economic agents. There is no reason to believe that economic agents will not adapt to the limit of their ability to protect their private assets and maximise utility (Fankhauser 1995b). The main problem is that at the moment there is no reliable way to estimate the amount of adaptation that will take place. In Chapter 5 an attempt is made to provide a solution.

1.4.2 Reasons for the relative paucity of climate change adaptation research

The treatment of adaptation as a serious policy option has been hampered by its relationship with mitigation and the political aspects of any promotion of adaptation as a policy option. In the very early days of climate change policy research at the international diplomatic level adaptation was seen as a 'dirty word'. 'Adaptation to a changing climate will be unavoidable. But it is a subject that carries a heavy ideological freight, for many people in the environmental movement suspect that any discussion of adaptation can only distract attention from the efforts to cut emissions' (Anderson 1997, p. 13). As time has gone by and more progress has been made towards the implementation of mitigation policy, the issue of adaptation has gained more attention and respect within diplomatic, environmental and academic groups.

There are several reasons why adaptation has not been studied nearly as much as mitigation as a viable climate change policy. The most important is that due to the precautionary principle it was considered in the planet's best interest that prevention/reduction would be better than a cure. Political factors have dominated since the climate change issue gained prominence. Kates (1997) explains that the main reason adaptation has been neglected as a climate change issue is that it has been dominated by two schools of thought that both discourage adaptation as a climate change issue. The first is what Kates calls the 'Preventionist' school where drastic action is advocated in the form of mitigation leaving the focus firmly off adaptation. The other school is described as the 'Adaptationists' who regard both adaptation and mitigation as not necessary as society should be able to adapt naturally and that any interference such as adaptation policy may cause higher social costs than climate change itself. Kates suggests that these two extreme views dominated until the 'Realist' school emerged after the IPCC SAR in 1995. Since then adaptation has held a much more significant place in the international climate change debate.

It is posited by Burton (1996) that any demonstration of the likely success of adaptation policies would substantially weaken the resolve of governments around the world to commit to legally binding mitigation targets. Burton states that the initial view of mitigation and adaptation policies being complementary issues changed to them being thought of as substitutes and subsequently claims for the need for large reductions in emissions were called for in the late 1980s. Research on adaptation was seen as a substitute to mitigation. Those studying adaptation options included the fossil fuel industry and oil exporting countries. Scientists became polarised between the issues and it became unpopular at an international level to be seen to advocate adaptation. It is claimed by Burton that this continued through to the Berlin FCCC conference in 1995. It is also apparent that this type of division has occurred at the national and state levels which has made it difficult for the formulation of climate change strategies that incorporate complementary adaptation and mitigation policy formulation.

1.5 Objectives of this book

In this section the objectives of this book are described to provide a sense of what the book is trying to achieve. The objectives of this book can be categorised as follows:

1. To provide an aggregate economic impact estimate for SEA, under $2\times CO_2$ conditions. Any further mention in this book of $2\times CO_2$ refers to the arbitrary standard point of measurement used for scientific and economic climate change studies where a level of twice pre-industrial atmospheric concentrations of carbon dioxide is used as a benchmark. While the benchmark has no particular significance for atmospheric chemistry it serves as a focal point that can be used for all disciplines to study climate change. An aggregate climate change impact estimate refers to the estimation of the total economic impact of the effects of climate change across all climate sensitive sectors of the economy.
2. To model the economic and environmental dynamics of climate change impacts for SEA by implementing a model based on the DICE (Nordhaus and Boyer 2000) framework: the South East Asia Dynamic Integrated Climate and Economy (SEADICE) or South East Asian DICE model.
3. To generate forecasting results from this model that will provide policy makers insight into the economic impact that climate change might have on major economic and environmental variables in SEA.
4. To explore adaptation to climate change as a concept and attempt to incorporate climate change adaptation into the dynamic optimal control economic model implemented in this book.

5. To suggest policy alternatives for the region based on the modelling results and arguments of the book.

1.6 Contributions of this book

The impacts of climate change are being felt already. As reported in IPCC (2001a) observed changes have occurred such as the shrinkage of glaciers, lengthening of mid-to-high latitude growing seasons and declines in plant and animal populations. This fact indicates that climate change is an important area for research for many disciplines, including economics. Consequently any contributions that can be made to the literature are also important. The contributions this book makes to the literature are as follows:

1. It is the first time economic impact estimates have been made for a variety of climate change sensitive sectors of SEA and aggregated to find the total impact of climate change under 2×CO_2 conditions for SEA.
2. For the first time an integrated optimal control model specifically representing the SEA region is implemented. Integrated models have so far mostly represented developed countries and regions. Representations of developing regions of this type are rare and to the author's knowledge no research of this type has been conducted for SEA, therefore it is a contribution to the literature.
3. A contribution is made by the novel application of the concept of adaptation to climate change to a dynamic optimal control model. This is done by applying the techniques of endogenous growth theory to the model implemented in this book. This is the first time to the author's knowledge that the level of technology has been assumed to determine autonomous adaptation and the application of endogenous technical progress has been used to represent this relationship in an optimal control modelling framework.
4. Based upon the arguments throughout the book and the findings of the SEADICE model, policy suggestions are made for both mitigation and adaptation. The policy recommendations by themselves are not controversial, however they represent a contribution because they are based on the unique modelling results of this book.

1.7 The structure of this book

Chapter 2 begins by defining the region covered in this book and describing its social, economic and environmental characteristics. The history of environmental awareness in the SEA region is explained. It becomes apparent from examining the GHG emission profile of the region that SEA is uniquely positioned. This is also apparent when the likely impacts of climate change for the region are explained. Economic growth and its relationship with the environment and in particular climate change are explored and it is found that for SEA, per capita emissions and emissions per unit of Gross Domestic Product (GDP) are still rising, whereas for many developed countries these values are now decreasing. Thus, the chapter sets up the geographical scope of the book and the climate change and environmental characteristics of the region. This provides the basis for the impact estimates made in Chapter 3.

In Chapter 3 an attempt is made to estimate the likely climate change impacts for SEA. With a review of the relevant literature it is found that this particular type of cost-benefit estimation is still developing and that most studies of this type have concentrated upon developed countries. Thus, an application to SEA provides a contribution to the literature. Estimates are made for several climate sensitive sectors for SEA. Significant impacts are predicted, particularly for the coastal and agricultural sectors of the economy. The sectoral results are combined and the aggregate result is found to be in excess of 5% of GDP for $2 \times CO_2$ climate change conditions. This indicates that economic output for SEA will be reduced by 5% as a result of the climate change conditions where global atmospheric concentrations of CO_2 are double the pre-industrial level. This result is used for Chapter 4, which implements the SEADICE model.

The climate change impacts found in the previous chapter provide data that contributes towards the implementation of a model that will enable the forecasting of important environmental and economic variables for the region. A review of Integrated Assessment Models (IAM) of climate change is provided along with a discussion of some of the important issues associated with climate change economic modelling. The SEADICE model, which is used as the basis of this book is explained along with arguments for and against the framework. The differences between the DICE model and the SEADICE model are explicitly detailed as well as the reasons behind the use of a spreadsheet program as the solution tool used for the model. The model results are then presented showing the dynamic paths of many economic and environmental variables for SEA. The SEADICE model and its forecasting results are a contribution to the literature as this type of modelling has not previously been applied to SEA. The SEADICE model serves as the basis for the application of adaptation in Chapter 5.

The concept of adaptation is analysed in Chapter 5 from a theoretical perspective, from broad scientific versions to how economics deals with the concept. A discussion follows on how the concept is different for economics compared to the natural sciences and that economics has had historical difficulties incorporating it into theory. The concept is then examined with respect to how it has been defined in terms of climate change. Like other disciplines, it is found that different definitions for climate change adaptation exist. Climate change adaptation is split into the concepts of autonomous and planned adaptation and it is explained that the estimation of planned adaptation is dependent upon the estimation of autonomous adaptation. This fact is largely ignored in the literature. It is suggested that the economic concept of endogenous technical change might be used to represent autonomous adaptation, given the theoretical discussion in the chapter and the assumption that the level of technology is a determinant of climate change vulnerability and hence autonomous adaptation. Endogenous technical progress is explained and a particular method using endogenous technical progress is implemented with the SEADICE model. After a review of other economic modelling assessments of climate change adaptation, results are presented from the endogenous version of the SEADICE model. The results indicate that the influence of endogenous technical progress and hence autonomous adaptation could be significant, with the consequence that impact studies could be substantially compromised if autonomous adaptation is not included in the calculations.

A move is then made from theory to policy in Chapter 6 with a discussion of the mitigation policy options available for climate change in SEA. Several policy options are examined with respect to their relevance for SEA including the flexibility mechanisms of the UNFCCC and no regrets possibilities. Two policy recommendations are made, firstly that SEA target mitigation efforts at participating in Clean Development Mechanism (CDM) activities with Annex I countries. Secondly, those no regrets policies that are most easy to identify should be implemented particularly in the forestry and energy sectors using Demand Side Management (DSM) techniques. Following these recommendations for mitigation policies, Chapter 7 provides recommendations for adaptation policies in the region.

Policy options specific to adaptation are presented in Chapter 7 including some arguments of why adaptation should be prioritised by the countries of SEA. Adaptation policy is defined at the start of the chapter followed by a review of the international rules governing adaptation policy. Reasons are provided for the enhanced profile of adaptation as a climate change policy. Different methodologies for the identification of adaptation policies are discussed as well as the barriers to identification that exist. It is concluded that significant opportunities exist for adaptation policies. To facilitate this it is suggested that a regional institution should be utilised to take advantage of pooled academic resources. Several arguments and findings from throughout the book are used to justify the recommendation that the resources of ASEAN are to be used as the best option for the identification of adaptation policies for the region.

2
Issues in Climate Change for South East Asia

2.1 Introduction

The environmental, social and economic problems associated with climate change present a unique policy challenge to every nation on earth. The general issues surrounding the economic effects of climate change have been examined for the whole of Asia by several authors (Bhattacharya, Pittock and Lucas 1994; Qureshi and Hobbie 1994; Amadore et al. 1996; Erda et al. 1996; Sanderson and Islam 2000a). However, less work has been focused upon the specific region of SEA.[8] In fact the only attempt the author could find at an overall analysis of the economic impact of climate change on SEA was Parry et al. (1992). In Parry et al. the effects of climate change were discussed but it was limited as a result of the paucity of empirical sectoral impact estimates for the region at the time. Precisely how climate change will impact upon SEA is yet to be determined, but it is known which regions of the world and sectors of individual economies will likely be most vulnerable (IPCC 2001b). The region itself is a mix of nations at various stages of development and with different political systems. However it is small enough geographically that there are many similarities in climate and therefore, ecosystems, agriculture, etc. between the countries of the region, where for the purposes of climate change policy, resources can be pooled to obtain improved outcomes for the entire region.

This chapter serves two main purposes, firstly it describes the economic and environmental aspects of SEA relevant for this book. Secondly, it describes the climate change characteristics of the region including the structure of emissions, both internally and globally and the rate and efficiency of emissions in an economic development context. This chapter provides the geographical scope and economic and climate change characteristics for SEA that provides support for arguments developed later with respect to model implementation and policy suggestions.

The specification of the climate change problem in regional terms is justified. Parson (1995) calls for economic studies to be more relevant for policy analysis by adjusting their spatial and sectoral resolution downwards. If more regional studies are to be done on the developing world, policy formulation will be more specific to those regions of the world deemed to be most vulnerable to climate change. Climate change impact assessments should be more focused on regional issues due to the significant differences in impact estimates between regions. While the United States and other developed nations are consistently forecast to suffer $2{\times}CO_2$ damage of between 1 and 2% of GDP (Nordhaus 1994a; Fankhauser 1995b), Tol (1996) estimates $2{\times}CO_2$ damage of 8.6% and 5.2% for South and SEA and Centrally Planned Asia respectively.

2.2 South East Asia

SEA is the most economically advanced and populous part of the tropics around the globe. These facts alone make the region important for study. In recent decades economic growth has been rapid due to factors such as the efficient use of new technology, substantial targeted public and private investment, a historically stable political environment and an increasingly skilled workforce (Dixon 1991). For the purposes of this book SEA is defined as consisting of the countries; Malaysia, Indonesia, Laos, Vietnam, Singapore, Philippines, Thailand, Myanmar, and Cambodia.

The historical recognition of SEA as a distinct entity by Western nations is fairly recent. Before the 1940s it was known as amongst others, Further India, Far Eastern Tropics or Indo-China. SEA has generally been recognised as a distinct region since the end of World War II. Since then, the main form of conflict has been political tensions between the capitalist and socialist nations in the region, which was in a way institutionalised in 1967 by the creation of the Association of South East Asian Nations (ASEAN) which consisted of the capitalist nations. This military-political divide originally limited the possibility for comprehensive economic cooperation throughout the region. However, these barriers are eroding over time; since ASEAN's initial membership of five (Indonesia, Malaysia, Singapore, Thailand and the Philippines), Brunei joined in 1984, Vietnam in 1995, and Laos and Myanmar in 1997, with Cambodia the final country to join the association in 1999.

2.2.1 Recent economic and social development in South East Asia

SEA has experienced a period of unprecedented development in recent decades (World Bank 2000; Sanderson and Islam 2001). Throughout the region many countries have experienced rapid development in the levels of many social and economic indicators. The extent of the development experienced in recent decades will be examined by looking at selected economic and social statistics. Recent economic development in SEA is characterised by substantial and rapid industrialisation, urbanisation, and structural, social and institutional transformation.

It is generally agreed that economic development in SEA has been caused by a combination of population growth, high rates of investment and savings and the strategic support of industries by governments of the region (Asian Development Bank 1991; Dixon 1991). Table 2.1 illustrates the excellent economic growth achieved

Table 2.1 Economic Growth for South East Asia 1978–97

	Average Annual Growth Rate of GDP (percent)	
	1978–87	1988–97
Indonesia	5.4	6.7
Cambodia		4.2
Laos		6.4
Malaysia	4.7	7.5
Philippines	1.2	3.0
Singapore	5.9	7.3
Thailand	5.1	6.8
Vietnam		6.9

Source: World Resources 2000–2001 Table EI.1.

throughout SEA during the period 1978–97. For 20 years very high growth rates (above 5%) were common throughout most of SEA. The rise of the economic 'tigers' of SEA has been very well documented throughout academic and journalistic literature along with the recent financial crisis, which significantly affected the region (Centre for Strategic Economic Studies 1998). In this respect this book has nothing new to add to the literature. The main purpose of this section is to provide some summary figures to illustrate the extent of recent economic development in the region and to provide a basis for later arguments related to SEA's climate change policy response.

SEA has not only developed economically but is also changing structurally and socially, as will be demonstrated by an examination of some structural social indicators. As can be seen from the indicators in Table 2.2, large changes have been experienced throughout most of SEA. The Organisation for Economic Cooperation and Development (OECD) average for all of these indicators is lower, demonstrating that not only has economic growth been more rapid in SEA but social and structural change is also changing rapidly. A most significant indicator of social structural change is the increasing urbanisation of SEA. From 1980–2000 it has been estimated that large movements of people have occurred in SEA. Throughout most

Table 2.2 Structural and Social Change in South East Asia

	Increase in Female Life Expectancy*	Movement of Labour Away from Agriculture**	Urbanisation from 1980 to 1999***
Developed	*3*	*4*	*4*
Philippines	9	12	25
Indonesia	13	11	23
Malaysia	7	18	23
Thailand	9	20	8
Vietnam	12	2	
Laos	10		13
Cambodia	22		4

Source: World Bank (2001).

* Years added to average life expectancy 1975–80 to 1995–00.

** % of labour force having moved away from agriculture.

*** % of population that has moved from country to urban regions 1980–2000.

of SEA almost one-sixth of the population has moved from agricultural areas to cities in only 20 years. The indicators are characteristic of a structural shift away from agricultural and subsistence dominated economies towards a more industrial based economic structure across SEA. However, it must be noted that this change has not been uniform. The large variations between countries can clearly be seen in Table 2.2. Such variety is also evident in the economic and political structure of SEA. While these examples are by no means exhaustive they serve to represent that recent social change in SEA has been relatively rapid.

SEA has been one of the most dynamic regions in the world in the last 30 years. Massive social and economic changes are taking place. The point that is being emphasised here is that the pace of change in SEA has been rapid and how it is dealt with is sometimes too great for the natural environment. In particular the rapid development of SEA has consequences for aspects of climate change contributions and vulnerability which will be explored later in the book.

2.2.2 Recent economic reform and growth

SEA was substantially affected by the Asian financial crisis in 1997 (Centre for Strategic Economic Studies 1998; Asian Development Bank 1999). However, the recovery since has been above expectations. Throughout 1999 all countries in the region experienced positive rates of economic growth, averaging a GDP growth rate for the region of 3.3% (Asian Development Bank 1999). The main reasons given by the Asian Development Bank (1999) for the strong recovery are an expansion of external demand and an improvement in commodity prices. In conjunction with reflationary fiscal and monetary policies across the region the recovery progressed considerably. Across the region economic growth rates have returned to pre crisis levels in recent years. Potential weaknesses such as the region's dependence on imported oil remain and keep the region susceptible to price fluctuations. Foreign capital has also returned since the crisis, reflecting renewed confidence in the political and economic stability of the region.

With respect to individual countries within the region, Cambodia has been undergoing reforms since the mid 1980s, however the economic structure is still largely agricultural. Cambodia has been the recipient of numerous international assistance measures which will determine much of the structural change in the nation in the short to medium term. Laos remains a largely agricultural economy significantly affected by neighbouring countries which import its raw materials, e.g., 40% of export earnings have come from timber during the last decade. Thailand is one of the economic success stories of SEA, strong economic growth has been the norm since the mid 1960s. Only the 1997 financial crisis has seriously threatened the growth prospects of Thailand. Three sectors; textiles, canned seafood, and electrical goods have accounted for the majority of the growth. While the number of people in poverty has reduced from around 33% to 10% in this time, it has become more concentrated within the rural sector, to the extent that 92% of all poverty is in this sector (Quibria 1995). Vietnam has experienced substantial change since the mid 1980s when the government initiated the Doi Moi in 1986,[9] where foreign policy and economic management practices were changed. While each of the countries of the region is at a different stage of economic development they mostly have in common a rapidity of change that is taking place. These examples of some of the economic reforms occurring throughout the region are typical of the transformations taking place. While each country may be starting from varied base levels and political ideologies, each has been moving in the same direction of further economic integration and market freedom.

2.2.3 The environment as an issue in South East Asia

The history of environmental awareness in SEA is worth examining if only to give a hint at the priority environmental issues have been given in society and government in SEA and as a background to the present climate change situation. For a comprehensive treatment of the subject see Grove, Damodaran and Sangwan (1998). Mishra, McNeely and Thorsell (1997) provide evidence for historical concern for the environment in SEA. The practice of protecting areas for their environmental qualities dates back to 684 AD in Indonesia where the king of Srivijaya established the first nature reserve on the island of Sumatra. The first national parks to be officially designated in SEA were Angkor Wat in Cambodia (1925), Taman Negara in Malaysia (1939), and Mount Arayat and Mount Roosevelt in the Philippines (1933). Since this time the amount of legally protected parks in SEA has grown to over 12% of the surface area of SEA (see Table 2.3).

Table 2.3 Protected Areas in South East Asia

Country	Land Area (sq km)	Protected Areas (sq km)	% Protected
Cambodia	181,000	34,022	18.8
Indonesia	1,919,445	321,087	16.7
Laos	236,725	24,400	10.3
Malaysia	332,965	110,222	33.1
Myanmar	678,030	9,725	1.4
Philippines	300,000	14,831	4.9
Singapore	616	32	5.2
Thailand	514,000	79,760	15.5
Vietnam	329,565	15,341	4.7
		Average	*12.28*

Source: World Conservation Monitoring Centre and IUCN (1994).

Since 1978 ASEAN has been actively involved in the environment and sustainable development issues in the region. A regular forum for senior environment government officials has been the ASEAN Senior Officials of the Environment (ASOEN) meetings since 1990, where several working groups have been developed to deal with issues related to the environment in SEA. The fourth strategic plan of 1994–98 developed by ASOEN included features such as responses to Agenda 21, harmonisation of environmental quality standards, government-private sector interactions, strengthening institutional and legal capacities to implement international environmental agreements, etc.

The countries of SEA are linked in many ways by their environments, the following example provides some insight as to the importance of these linkages. In mainland SEA, Thailand is the dominant economy. Since 1989 logging has been banned in Thailand. However, demand for timber products has still increased rapidly. Since 1989 Thailand has relied more on its neighbours for resources. The potential exists that the removal of unsustainable environmental practices in Thailand have now shifted demand for wood from within Thailand to the next cheapest available source of supply such as Myanmar, Laos and Vietnam thereby endangering their natural resource stocks. This type of relationship is not unique to forestry and is leading to an internationalisation of many environmental issues within the region of SEA. This change of perception and the way in which policy makers react to environmental consequences has important implications for other environmental problems such as climate change.

2.2.3.1 South East Asia's activities in international environmental law

Another perspective can be gained upon the seriousness with which a region takes environmental issues by examining its activities in international environmental law. A review of the region's activity with respect to this matter is provided in this section.

With respect to individual countries in the region, the record on environmental laws is as follows. Many of the environmental initiatives in the region came after the 1992 Rio Earth Summit where the issues of sustainable development and environmental conservation gained global attention. In Cambodia Article 59 of the 1993 Constitution was the only source of environmental regulation in the country. In 1996 an environmental law entitled Law on Environmental Protection and Natural Resources Management was legislated (Asian Development Bank 2000). In addition, several other laws are also being considered in relation to the environment. In Vietnam, the 1993 Law on Environmental Protection provides wide ranging commitments to the environment by the government. In 1993 the National Environment Agency was established. Laos has been undergoing economic reform since 1986 when the transition to a market based economy began. In 1992 the government established within the Prime Minister's office an agency responsible for environmental issues called the Science, Technology and Environment Organization. Its role is to develop a comprehensive national environmental policy framework encompassing compliance monitoring, management processes, dispute resolution and research on conservation. Indonesia's first Minister of the Environment was established in 1982; the 1994–99 five year plan highlighted the need for significant attention to environmental issues. Malaysia initiated the Environmental Quality Act in 1974 that established coordinated mechanisms to address environmental issues. Malaysia has also played a significant diplomatic role in the region; initiating both the 1989 Langkawi Declaration and the 1992 Kuala Lumpur Declaration. The Philippines has since 1992 initiated some substantial environmental reforms such as the creation of the Council for

Sustainable Development. They have also initiated a system where environmental accounting is now an institutionalised part of its official statistical collection and analysis system (Republic of the Philippines 1997; Bartelmus 1999). Singapore has maintained a strong government objective of being 'clean and green' since the 1960s. Subsequently, the environment ministry was established there in 1972. Thailand introduced its Enhancement and Conservation of National Environmental Quality Act in 1992. As can be seen from Table 2.4 the countries of SEA have been active in international environmental conventions since the 1950s.

> While the countries of the Mekong Region have signed a significant number of global and regional environmental conventions, the implementation of environmental conventions is relatively weak in most of them. However, in each country, capacity building programmes (Thailand and Vietnam in particular at this stage) are beginning to address the inadequacies of environmental legislation and attendant administrative and enforcement structures. (Boer, Ramsey and Rothwell 1998, p. 203).

While the regulation and enforcement of environmental policies and legislation are not yet as advanced as the developed world, throughout the region governments and communities are moving in the direction of enhancing their environment. This indicates the seriousness with which the countries of SEA consider domestic and international environmental problems and provides a pointer to the likelihood of the participation of the region in tackling global environmental problems such as climate change.

2.2.4 Climate change activities in South East Asia

While the countries of SEA are not part of the Annex I group of countries, which are expected to be the first to implement climate change policy, they are far from inactive with respect to climate change research.

An important event for the region was the Asia Pacific Leaders' Conference on Climate Change, which resulted in what has become known as the Manila Declaration. The conference took place in February 1995 in Manila, in the Philippines, where 250 delegates attended from 33 countries along with many significant dignitaries.

Table 2.4 Environmental Conventions to which South East Asian Countries are Party

International Convention	Indonesia	Malaysia	Philippines	Singapore	Thailand	Cambodia	Vietnam	Laos
International Convention for the Regulation of Whaling				✓				
FAO International Plant Protection Convention	✓	✓	✓		✓	✓		✓
Convention on the High Seas	✓	✓			✓	✓		
Treaty Banning Nuclear Weapon Tests in the Atmosphere, in Outer Space and Under Water	✓	✓	✓	✓	✓			✓
Vienna Convention on Civil Liability for Nuclear Damage			✓					
Treaty on Principles Governing the Activity of States in the Exploration and use of Outer Space Including the Moon and Other Celestial Bodies	✓	✓	✓	✓			✓	
Convention of the Wetlands of International Importance Especially as Waterfowl Habitat	✓	✓					✓	
Convention Concerning the Protection of the World Cultural and Natural Heritage		✓	✓		✓	✓	✓	✓

Table 2.4 Environmental Conventions to which South East Asian Countries are Party – *continued*

International Convention	Indonesia	Malaysia	Philippines	Singapore	Thailand	Cambodia	Vietnam	Laos
Convention on the Prevention of Marine Pollution by Dumping of Wastes and Other Matters	✓		✓	✓	✓			
Convention on International Trade in Endangered Species of Wild Fauna and Flora (CITES)		✓	✓	✓		✓	✓	
Convention on the Conservation of Migratory Species of Wild Animals			✓					
United Nations Convention on the Law of the Sea	✓	✓	✓	✓	✓	✓	✓	✓
International Tropical Timber Agreement	✓	✓	✓		✓			
Vienna Convention for the Protection of the Ozone Layer	✓	✓	✓	✓	✓		✓	
Convention on Early Notification of a Nuclear Accident	✓	✓	✓	✓	✓		✓	
Convention on Assistance in the Case of a Nuclear Accident or Radiological Emergency	✓	✓	✓		✓		✓	
Montreal Protocol on Substances that Deplete the Ozone Layer	✓	✓	✓	✓	✓		✓	

Table 2.4 Environmental Conventions to which South East Asian Countries are Party – *continued*

International Convention	Indonesia	Malaysia	Philippines	Singapore	Thailand	Cambodia	Vietnam	Laos
Basle Convention on the Control of Transboundary Movements of Hazardous Wastes and Their Disposal	✓	✓	✓		✓		✓	
Framework Convention on Climate Change (FCCC)	✓	✓	✓	✓	✓	✓	✓	✓
Convention on Biological Diversity	✓	✓	✓	✓	✓	✓	✓	✓
Convention to Combat Desertification in those Countries Experiencing Serious Drought and/or Desertification	✓		✓			✓		

Source: Boer, Ramsey and Rothwell (1998).

The Manila declaration, which was a declaration of the attendees of the conference, not a formal declaration of states, declared the position of the Asia Pacific countries on matters related to climate change. This declaration was presented to COP1 and demonstrated the commitment of the Asia Pacific region to the climate change debate. An important outcome of the Manila Declaration was the Regional Action Plan for Climate Change in the Asia Pacific. This action plan has three main objectives for action; (1) national/regional measures for scientific and technical advice and public education; (2) national/regional measures for adaptation to climate change impacts and vulnerabilities and (3) national/regional measures for mitigation of anthropogenic GHG emissions.

A major source of funding for climate change projects in SEA comes from the Global Environment Facility (GEF) which was established to foster international cooperation and finance actions to address four threats to the global environment: biodiversity loss, climate change, degradation of international waters, and ozone depletion. The GEF provides funding to developing nations for projects to enable compliance with COP directives such as the preparation of GHG inventories. This was accomplished with the GEF sponsored completion of the National Communications to the UNFCCC for most of SEA recently (Philippine's Initial National Communication on Climate Change 1999; Lao People Democratic Republic 2000; Office of Environmental Policy and Planning 2000; Singapore Ministry of the Environment 2000; Ministry of Science Technology and the Environment Malaysia 2000; Sugandy et al. 2000).[10] As of May 2000 $US72 million has been approved for such activities in 132 countries. As a reflection of the importance of SEA to the global climate change problem, 9.86% of total GEF funding was allocated to national projects in the ASEAN region (Global Environment Facility 1996; 2000). These projects are implemented mostly by the United Nations Development Programme (UNDP) and to a lesser extent by UNEP and the World Bank. Under the GEF guidelines enabling activities 'include [GHG] inventories, compilation of information, policy analysis, and strategies and action plans. They either are a means of fulfilling essential communication requirements to the Convention, provide a basic and essential level of information to enable policy and strategic decisions to be made, or assist planning that identifies priority activities within a country.'

(Global Environment Facility 1996, p. 9). Recently, the GEF undertook a comprehensive review of all of its enabling activities (Global Environment Facility 2000). The final conclusion of the GEF review was that 'support provided by the GEF for climate change enabling activities has substantially contributed towards assisting non-Annex I Parties in meeting their communication commitments under the UNFCCC' (Global Environment Facility 2000, p. 3). Therefore, the GEF is satisfied that at this stage the reporting requirements such as emission inventories have been progressing sufficiently for non-Annex I countries, including those in SEA.

Political support for climate change issues has been strong from the Philippines with the establishment of the International Agency Committee for Climate Change by Presidential order, which has involved high level representatives. In Vietnam it is the responsibility of the Hydrometeorological Service (HMS) for climate change issues and implementing programs for the purpose of fulfilling objectives of the UNFCCC.

There are also some internationally sponsored programs that are designed to facilitate climate change awareness and policy development in SEA. These include:

1. Asia Least Cost Greenhouse Gas Abatement Strategy (ALGAS), which covered 12 Asian countries and provided least cost mitigation options available for the specific countries involved.
2. Regional Studies on Global Environmental Issues, which is an Asian Development Bank (ADB) sponsored project, and provides policy options and estimates for the socioeconomic impacts of climate change for eight Asian countries.

As well as Government and Non Government Organisation (NGO) activities in climate change there is also a developing academic network in the region whose focus is climate change issues. The three major centres are as follows: the Impacts Centre for South East Asia (IC-SEA) based in Indonesia; the Asia Pacific Network for Global Change Research (APN) which is an inter governmental network of 21 countries supporting research activities on issues of global change effecting the region including climate change; and finally there is also the South East Asia START, the Global Change

SysTem for Analysis, Research and Training. The latter is a global network that encourages multidisciplinary research on the effects of global changes on the interactions of human and environment systems (Lebel and Steffen 1998). All of these institutions are promoting research activities, running conferences and strengthening academic networks which are providing substantially more scientific evidence of the possible impacts of climate change on SEA. As a result, papers are being produced in many areas related to climate change and its impacts for SEA. This includes important collected works such as those found in Qureshi and Hobbie (1994), Bhattacharya, Pittock and Lucas (1994), Chou (1994) and Amadore et al. (1996); crop production (Iglesias Erda and Rosenzweig (1996); land use and biodiversity (Lebel and Murdiyarso 1998); climate scenarios (Whetton 1996); economic impacts (Sanderson and Islam 2000b) and assessment methods (Jakeman and Pittock 1994).

The modelling of future climate change and variability for SEA has also been the focus of some recent studies. SEA is highly dependent on the timing and strength of the monsoon for sectors such as water resources, human life, agriculture and ecosystems. This factor provides an additional reason over and above that of standard changes in climate justifying the use of more focused weather models for the region. McGregor, Katzfey and Nguyen (1998) used the Division of Atmospheric Research Limited Area Model (DARLAM) to make accurate predictions of current climate patterns in SEA as well as predictions for precipitation and other climate variables. This type of data will be invaluable for future studies on the economic impacts of climate change that can incorporate SEA only climate data. When the future research plans (such as those outlined in Lebel and Steffen (1998)), are realised, the raw data needed for a book such as this will be significantly improved.

The main conclusion of this section can be summarised as follows; while SEA is still a developing region it is developing rapidly and this pace of development has consequences for both the causes and consequences of climate change for SEA. It has a long history of environmental awareness and these issues, in particular climate change, are becoming a more significant part of academic and international/local government policy making in the region.

2.3 Greenhouse gas emissions of South East Asia

2.3.1 The contribution of deforestation to climate change

Deforestation has been a high profile environmental issue throughout SEA for some time (Asian Development Bank 2000). Forests are vital for many reasons, including protection of watersheds, biological diversity, the maintenance of hydrological cycles and soil stabilisation. Economic development in the region has resulted in demand pressures that have been increasing for several contributing factors of deforestation. These include demand for agricultural and residential land, the domestic use of timber for construction and fuel and international demands for timber from resource depleted regions (Bautista 1990; Boonpragob and Santisirisomboon 1996). The severity of the deforestation problem is illustrated by the fact that the Philippines originally contained 16.5 million hectares (Ha) of forest, which has been reduced to only 5 million Ha (Cameron 1996). Between 1990 and 1995 the annual deforestation rate for SEA was 1.67%. This compares poorly with a global rate in the same period of 0.32% (World Resources Institute 2001). While deforestation is an environmental problem which has consequences for issues such as salinity, erosion and biodiversity, it also has impacts on climate change. Deforestation is one of the major contributors to climate change in Asia accounting for approximately 20% of CO_2 emissions (Sharma 1994). Within the region of SEA it is of even greater importance particularly for Thailand, Indonesia, Malaysia and the Philippines. For example, a GHG emissions inventory for Indonesia revealed that deforestation and land use changes account for 78% of total CO_2 emissions (Qureshi and Hobbie 1994).[11] The proportions of total CO_2 equivalent emissions for other countries is as follows, Vietnam 28%, Thailand 35%, Philippines 50% (ALGAS 1998b, 1998c, 1998d).

Currently no accurate data exists for the emissions of GHG as a result of land use change and deforestation for SEA, however institutions such as the IC-SEA are working towards this goal (Lebel and Murdiyarso 1998). The data that will become available in the future from further research will be significant. However, before the data can be applied to models such as SEADICE research is needed into measuring in an accurate way how land use has already changed in the region. At this stage it can only be emphasised that the issue is a

Table 2.5 Change in Energy Consumption and Production for Selected South East Asian Countries

	Energy Consumption and Production	
	% Change in Total Production from 1987 to 1997	% Change in Total Consumption from 1987 to 1997
Cambodia	–	–
Indonesia	236	71
Malaysia	76	161
Philippines	241	55
Singapore	126	202
Thailand	225	157
Vietnam	216	68

serious one for SEA due to the large amount of tropical rainforest in the region and the historically high levels of deforestation recorded in recent decades.

2.3.2 Energy use in South East Asia as a contributor to climate change

The role of energy use in SEA, as a contributor to climate change, is also becoming increasingly important (NISTEP 1991; Malik 1994; Sharma 1994). SEA has experienced some of the highest growth rates in demand for energy of any region on earth in recent decades. It is apparent from Table 2.5 that both production and consumption of energy have increased dramatically in the SEA region in recent decades. The rapid increases in consumption of energy have been driven by the development factors discussed earlier such as population growth, urbanisation (and the resulting higher commercial energy use lifestyle), and economic expansion of industry and commercial agriculture. There is still time to implement emission saving technologies that allow for the further introduction of infrastructure, and unhindered growth. This is discussed later in Chapter 6.

2.3.3 Overall structure of emissions

As Table 2.6 demonstrates, the structure of CO_2 emissions from SEA is comparable to South America where a relatively small proportion

Table 2.6 Structure of Regional CO_2 Emissions

	% of Total CO_2 Emissions 1996				
	Coal	Petrol	Gas	Gas Flaring	Cement Manufacturing
World	38%	40%	18%	1%	3%
Asia	57%	30%	7%	0%	6%
Europe	35%	34%	29%	1%	2%
Middle East and North Africa	7%	54%	28%	5%	5%
Sub-Saharan Africa	52%	33%	3%	10%	2%
North America	35%	40%	24%	0%	1%
Central America and Caribbean	4%	75%	17%	1%	3%
South America	10%	63%	19%	3%	5%
Oceania	58%	27%	14%	0%	1%
South East Asia	16%	58%	19%	1%	6%

Source: World Resources Institute (2001).
Note: Rounding may result in totals being out by plus or minus one percent.

of emissions come from coal based activities. While this only represents CO_2 and not the other GHGs, CO_2 is the most important GHG and represents greater than half of all emissions. The implications of this particular structure of emissions will be discussed later in the chapter.

2.3.4 South East Asia's place in the global greenhouse

How is SEA represented in the climate change debate as a region in terms of economic size, vulnerability and as a polluter? As a region SEA's share of global CO_2 emissions has grown from 1.3% in 1980 to 3.1% in 1996, while during the same period SEA's share of total Asian CO_2 emissions grew from 10.4% to 12.5%. This illustrates that emission growth rates for SEA are much higher than global and even Asian averages and that regionally SEA is closely linked to the greater Asia region with respect to emission growth rates. Figure 2.1 reveals that Thailand, Indonesia and Malaysia are the greatest emitters of CO_2 in the SEA region. Despite the high growth rate of overall emissions it is expected that emissions per capita will still be lower relative to the developed world. However, that situation is changing over time. According to Figure 2.2 per capita emissions

Figure 2.1 Individual Countries' Share of Total South East Asia Carbon Dioxide Emissions 1996

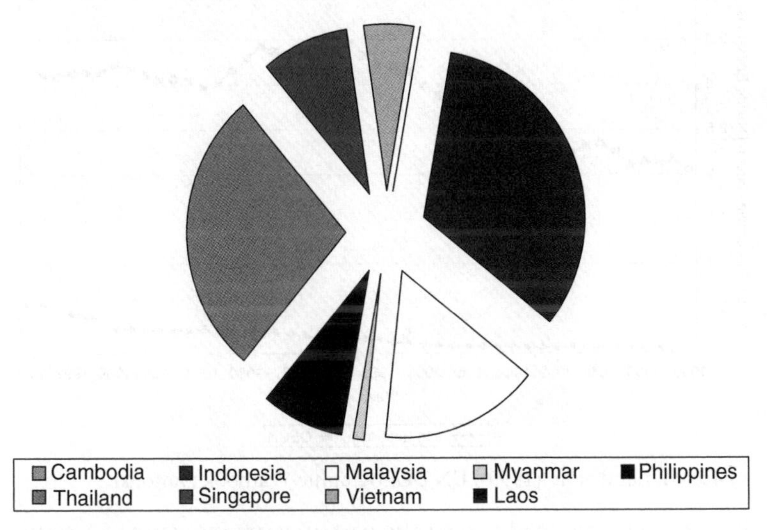

Source: Marland et al. (1999).

Figure 2.2 South East Asia Per Capita Carbon Dioxide Emissions

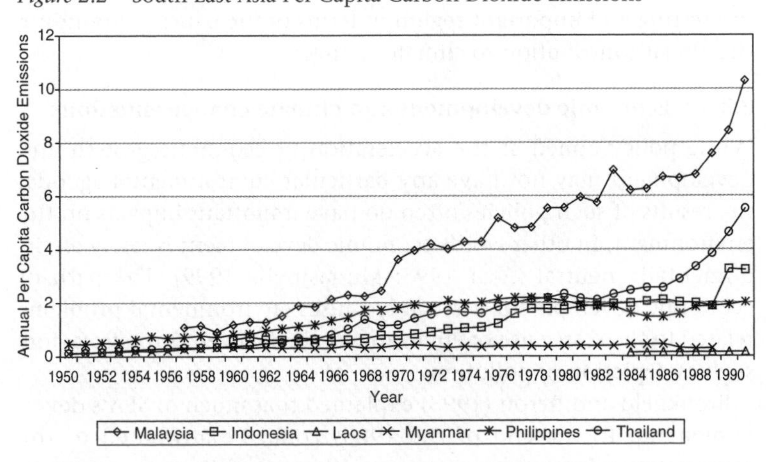

Source: Marland et al. (1999); UN Demographic Yearbook (various).

Figure 2.3 Annual Per Capita CO_2 Emissions 1950–95 (tons)

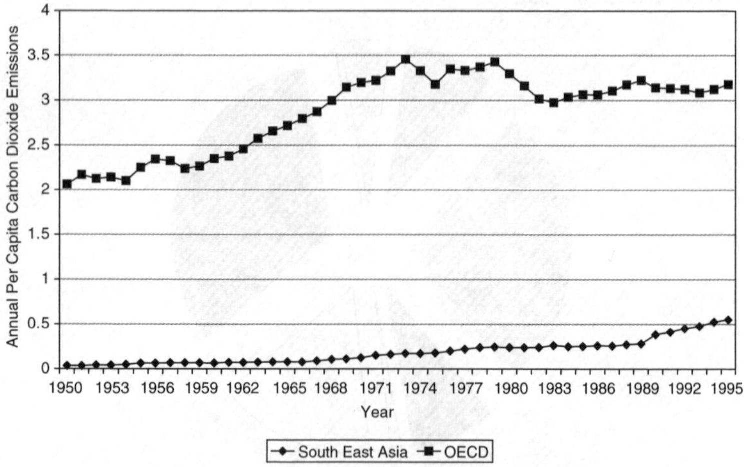

Source: Marland et al. (1999); UN Demographic Yearbook (various).

have been rising steadily over the past 40 years for many SEA countries. The developed world emits about 13.7 tonnes of CO_2 per capita annually (Turton and Hamilton 1999). While countries such as Thailand and Malaysia are fast approaching those levels the region as a whole still has a long way to catch up to the developed world as can be seen in Figure 2.3. These facts present SEA as a very interesting and important region in terms of the issues surrounding the global contribution to climate change.

2.3.5 Economic development and climate change emissions

While policy aimed at the acceleration of economic growth and development may not have any particular environmental agenda; the results of such policies often do have important impacts on the environment. In other words economic development is rarely environmentally neutral (Azar 1995; Munasinghe 1999). The path of economic development inevitably causes environmental problems related to the scarce use of environmental resources, through factors such as depletion or pollution.

Brookfield and Byron (1993) explained that much of SEA's development up to the end of the 1980s could be attributed to the exploitation of natural resource wealth, and while some advancements in industrialisation had been made as a result of governmental industry support there remains a significant dependence on primary resource exploitation and/or foreign capital. At the macro level Munasinghe (1999) argues that the stability of wages, prices and employment can be useful for the environment. The argument is that as these types of variables become more stable, firms and households tend to take longer term views which are compatible with environmentally sustainable activities. It is this type of argument which underlies the Environmental Kuznets Curve (EKC). The EKC[12] is an inverted U-shaped curve showing the relationship between a certain pollutant and economic development. In the case of global warming, the relationship would be between the emissions of CO_2 and GDP. The basic theory is that as an economy industrialises and incomes increase pollution also increases,[13] but at some level of income the level of pollution starts to decrease as individuals become more concerned about the environment and have more disposable income to spend on environmentally friendly products, etc. Another possible reason for the reduction in pollution is the structural shift from an industrialised economy to a more information and services based economy. It is only when consumers have the leisure time and the income to prioritise the improvement in the human environment that some environmental aspects start to be improved.

The main limitations of the EKC are as follows (List and Gallet 1999):

1. The turning point may be too high for it to be practical for developing countries or for it to prevent serious environmental damage.
2. The significance of EKC has been shown to reduce when additional variables are included in its calculation.
3. As with many areas, a lack of available data restricts the strength of analysis.
4. Not all pollutants have been shown to exhibit EKC properties.

Limitations such as these, however, do not prevent the examination of the insights the EKC provides for climate change in SEA. Figure 2.4 shows the relationship between per capita CO_2 emissions and per capita GDP for Indonesia, Malaysia, Philippines and Thailand.[14] The relationship shows that as these economies have

Figure 2.4 Per Capita Emissions and Per Capita GDP Relationship (South East Asia)

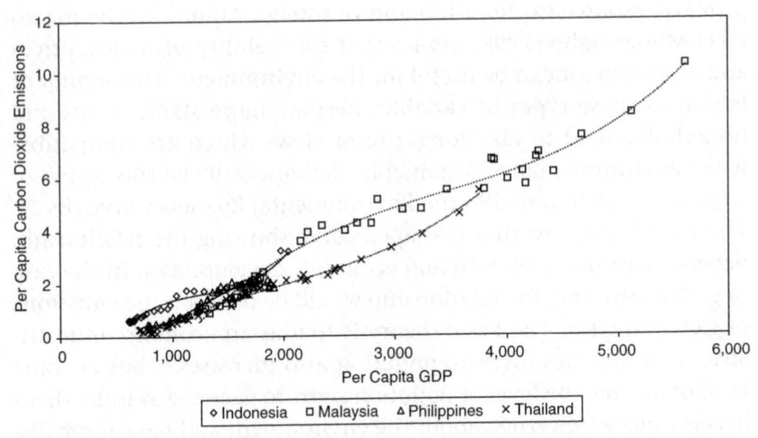

Source: Data compiled from Marland et al. (1999); Heston and Summers (1995).

Figure 2.5 Per Capita Emissions and Per Capita GDP Relationship (Developed Countries)

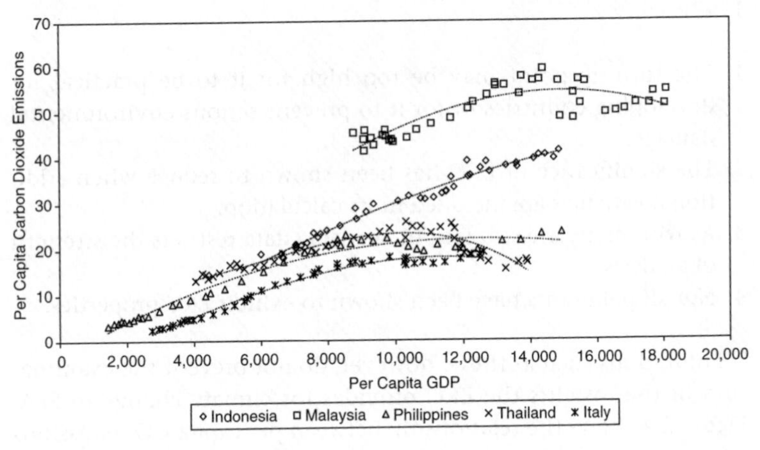

Source: Data compiled from Marland et al. (1999); Heston and Summers (1995).

been developing and income has been rising, emissions have also been rising, suggesting a positive relationship between these two variables. This indicates that the countries of SEA are becoming more CO_2 intensive and are yet to reach the potential 'peak' of the EKC curve where the economy becomes less CO_2 intensive. This contrasts with Figure 2.5, which shows the same relationship for the developed countries of Australia, United States, France, Japan and Italy. Most of these countries seem to be reaching the peak of the relationship, indicating that as per capita GDP increases past a certain point emissions per capita plateau and seem to begin to fall.

Figures 2.6 and 2.7 present a similar account to Figures 2.4 and 2.5. Figures 2.6 and 2.7 show the relationship between CO_2 emissions per unit of GDP and per capita GDP. CO_2 emissions per unit of GDP represents the efficiency with which an economy produces CO_2 emissions. Therefore, those economies with a high CO_2 per unit of GDP are inefficient with respect to their production of goods and services and consequently CO_2 emissions. It can clearly be seen that the developed countries are becoming more efficient per unit of GDP whereas the developing countries of SEA are still in the development phase where CO_2 emissions per unit of GDP are rising as the economy develops.

Figure 2.6 South East Asia Emissions Efficiency

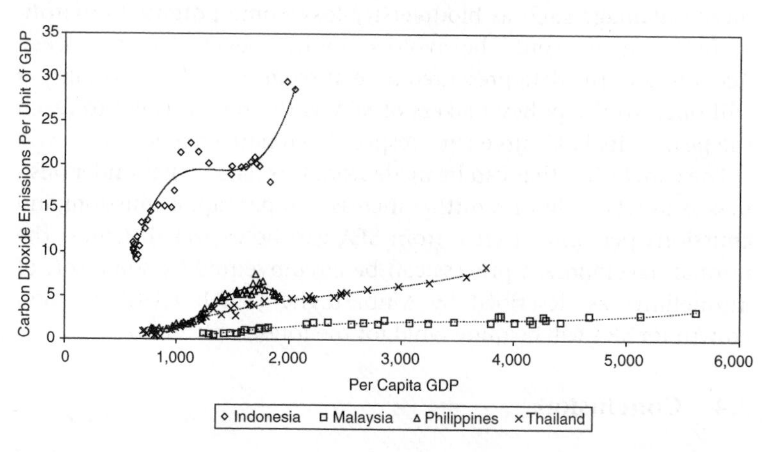

Source: Data compiled from Marland et al. (1999); Heston and Summers (1995).

Figure 2.7 Developed Countries Emissions Efficiency

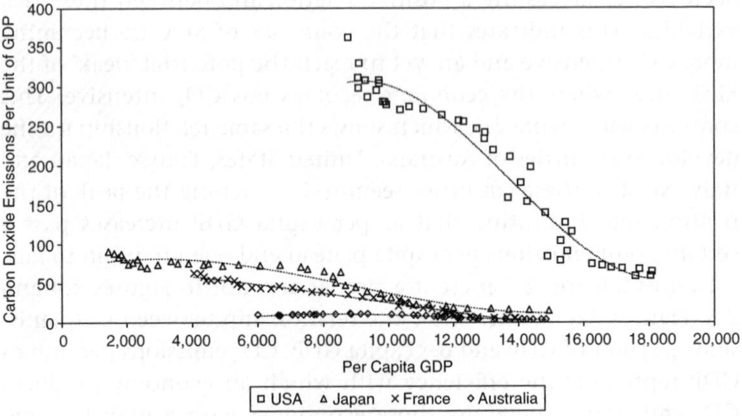

Source: Data compiled from Marland et al. (1999); Heston and Summers (1995).

Munasinghe (1999) argues that it may be possible for developing countries to 'tunnel' through the EKC. In other words, by introducing policies that are de-linked with respect to environmental degradation and economic growth. So, instead of reaching the normal peak of the EKC with the possibility of some irreversible environmental damage such as biodiversity loss, some potential environmental damage could be avoided in the development process. According to the data presented here, it seems that this possibility is still open to the policy makers of SEA as the region is yet to reach the peak of its EKC curve with respect to climate change.

The conclusion that can be made from this data is that under business as usual conditions, further increases in per capita emissions and emissions per unit of GDP from SEA can be expected. Unless the normal development process can be circumvented by some sort of 'tunnelling' as described by Munasinghe (1999) GHG emission growth for SEA will be quite rapid for the foreseeable future.

2.4 Conclusion

The purpose of this chapter is to provide the geographic scope of the book, and to look at some of the special environmental and climate change characteristics of the region. SEA has been developing rapidly in recent decades, this presents both problems and opportunities for climate change issues. Currently, the countries of SEA are still in the phase of economic development where per capita and per GDP pollution is increasing. At this point in time the detailed emission profile of the region is not known as the type of detailed emission inventories that the developed world has now completed has not been done for the developing world. However, it can be surmised that two of the most important sectors are forestry/deforestation and energy. These sectors, in particular forestry/deforestation have many similar characteristics across the region. Although on a global scale the emissions from this region are relatively small now, they are growing at rates higher than the global average and therefore are becoming a greater share of global emissions. It can be concluded from this chapter that the region of SEA is an interesting and important one for climate change economics. While this chapter has primarily focused upon the characteristics of SEA with respect to its emissions the next chapter will focus upon the possible impacts of climate change on SEA and how the economic and social structure discussed in this chapter influences SEA's vulnerability to the effects of climate change.

3
Climate Change Impact Estimates for South East Asia

3.1 Introduction

This chapter leads on from the discussions in Chapter 2 of the economic and climate change emission structure of SEA to make an aggregate estimate for the impact[15] of $2 \times CO_2$ climate change for the region. It is based upon and expands on the preliminary estimates made for SEA in Sanderson and Islam (2001). As explained in Section 1.5, a $2 \times CO_2$ climate change aggregate impact estimate refers to a prediction of the extent of the economic effects of climate change given in terms of aggregate GDP for the future situation of the benchmark level of double pre-industrial atmospheric CO_2 concentrations. Others have estimated the equivalent impacts for particular countries within SEA (Qureshi and Hobbie 1994) and have distinguished the SEA region as part of global estimates (Tol 1996). However, this is the first example (to the author's knowledge) of an aggregate $2 \times CO_2$ climate impact assessment for the region of SEA which uses data specifically related to the region and not extrapolated solely from other regions.

3.2 Aggregate climate change impact estimation methods and results

3.2.1 The methodology

Aggregate climate change impact estimates at the national and global level have been made for the last decade (Nordhaus 1991, 1994a; Cline 1992; Fankhauser 1995b; Tol 1995, 1996; Mendelsohn et al. 2000; Nordhaus and Boyer 2000) and continue to be made despite criticisms of the worth of the exercise (Schneider 1997; Morgan et al. 1999). The methodology used in this chapter follows that used in Nordhaus and Boyer (2000) and involves either making estimates of climate change impact in climate sensitive sectors or using the figures adopted by other authors. These estimates are all monetised in terms of a percentage of GDP at a common level (in this case $2 \times CO_2$). The $2 \times CO_2$ level of atmospheric concentrations has been used as a benchmark for both scientific and economic studies of the effects of climate change. This level is arbitrary (although it is a goal of institutions such as the IPCC and UNFCCC to keep concentrations below such a level) and is used mainly as a point of reference for researchers. There is no reason to assume that concentrations of CO_2 will not exceed the $2 \times CO_2$ level. In fact that scenario could occur as early as the year 2050 (Fankhauser 1995b). Therefore, it is always prudent for authors to make the point that whatever estimates are made, they may be much different if the $2 \times CO_2$ barrier is passed (Cline 1992). Unknown threshold points with catastrophic consequences are always a possibility as little is known of climate science.

To gain the overall impact figure, each of the sectoral estimates is summed, hence the method is often referred to as *enumerative* (Fankhauser 1995b). As mentioned earlier, the data collection and impact estimates in this chapter follow closely the methods employed by Nordhaus and Boyer (2000). The main reasons for following Nordhaus and Boyer (2000) are that the estimates made by Nordhaus and Boyer are specifically made for the DICE model upon which the SEADICE model of this book is based. Also, the regional groupings of some of the sectoral estimates are taken from Nordhaus and Boyer (2000) as they are the closest proxy for SEA that is available. The reader must be cautioned that many of the estimates presented here are still 'fairly ad hoc' as described by Fankhauser (1995b, p. 56). Even those estimates based on state of the art research are qualified due to the high levels of uncertainty surrounding the exact nature of climate change. Many of the sectoral estimates are made using the Willingness To Pay (WTP) criteria. The WTP approach estimates the amount of money that society is willing to pay to prevent climate change and its associated impacts. According to Nordhaus and Boyer (2000) the advantage of this method is that it can utilise different approaches to measuring impacts. This method has been criticised on equity grounds by Fankhauser, Tol and Pearce (1997). They argue that giving differentiated values according to income for identical goods is unethical, particularly in the valuation of mortality risk. The author acknowledges these limitations, however given the uncertainties already present in this type of analysis, at this stage the priority should be to provide results that are transparent so that further research is enabled in this area for a wider range of academics, particularly in the SEA region.

3.2.2 Other aggregate impact estimates

As mentioned in the previous section several major attempts have been made to estimate the potential overall impact of climate change. So far the vast majority of studies have been on developed nations and the United States in particular. This is mainly due to the relatively larger quantity of high quality scientific climate change analysis that has already been done on the United States. Economic research in the area of aggregate climate change impacts is driven by the quality and quantity of the scientific information available. So far, the developing countries have been grossly underrepresented with respect to estimates of the ecological, social and economic impacts of climate change. However, studies have been completed for some countries in Asia (Matsuoka, Kainuma and Morita 1995; Jia 1996). There are still numerous gaps in the literature with regards to both coverage within sectors of countries and between countries. Over time this situation is changing as international organisations, and developing countries themselves tackle the problem. In this section some of the most important aggregate climate change impact studies will be reviewed.

The first attempt at an aggregate estimate of the economic impact of climate change was made by Nordhaus (1991) where a global estimate of 1% of GDP was made for the impact of climate change which was primarily reliant on extrapolating United States data to represent global values. This was followed by a comprehensive study by Cline (1992) who used cost-benefit analysis to project the potential for long term climate change impact for the United States. Cline's work is noteworthy for its emphasis on the very long term potentials of climate change impacts. It highlighted the dangers of concentrating only on the standard 2×CO_2 estimates. Nordhaus (1994a) extrapolated data for the United States again to make a prediction for the global impact of climate change this time using the DICE model. This was the first impact estimate made using a dynamic economic modelling framework. As the SEADICE model implemented in this book is based on the DICE model, more comprehensive analysis can be found in Chapters 4 and 5. Fankhauser (1995b) made global damage estimates as well but went beyond previous studies by also making estimates for five geopolitical regions (European Union, United States, countries of the former USSR, China and OECD). He too found global climate change impacts to be within the 1–2% range of earlier studies. One of the most recent studies is Mendelsohn et al. (2000) who estimated the impact resulting from climate change for all countries using the Global Impact Model (GIM), which they claim features the economic rigour of a top down type model and the spatial detail of a bottom up model. The limitations cited by the authors such as impact functions being based on United States data applied to the world; shortage of non-climate data for each country; exclusion of non-market effects; spatial resolution difficulties with respect to different sized countries; lack of dynamics and the absence of the effects of sulphate aerosols, are by no means exclusive to this model. The results of the GIM model indicate that climate change benefits may be obtained by cooler climate countries (those in high latitudes) while those in warmer regions will be the most negatively effected. Generally the damage estimates appear to be somewhat smaller than other studies, for example the model finds global climate change benefits for all scenarios up to 3.5°C, although very small (ranging from 0.02–0.16% of GDP). However Mendelsohn et al. (2000) does find that climate change will be most detrimental to tropical regions which is relevant for this book.

The DICE-CHN model of Jia (1996) is the closest equivalent to the SEADICE model presented in this book both in terms of structure and geographical specificity. In this model China is examined within a DICE type framework while Rest Of the World (ROW) emissions are treated as exogenous to the model. The model is also distinctive by the fact that output for China is determined exogenously. This is done according to the author as a result of the paucity of data available. In this model the only value used for the impact function coefficient was an estimate of 5% impact on agriculture for $2\times CO_2$ conditions. No other sectoral impact estimates were included. Consequently, overall impact were found to be less than 1% of income.

From Table 3.1 the results from the studies discussed previously can be seen. The most striking observation is that while most of the studies shown have comparable overall results (ranging from 1–2.5% of GDP) the range of sectoral estimates varies substantially. Therefore, while a consensus of sorts exists for the overall estimation of impacts of $2\times CO_2$ climate change impact, each author has taken different paths to the same conclusion. Fankhauser (1995b)

Table 3.1 Climate Change Impact Estimates for the United States (1990 $US Billions)

	Cline (2.5°C)	Fankhauser (2.5°C)	Nordhaus (3°C)	Titus (4°C)	Mendelsohn* (2°C)	Tol (2.5°C)
Agriculture	17.5	3.4	1.1	1.2	50.0	10.0
Forest Loss	3.3	0.7		43.6	9.0	
Species Loss	4.0	1.4				5.0
Sea Level Rise	7.0	9.0	12.2	5.7	0.0	8.5
Electricity	11.2	7.9	1.1	5.6	1.0	
Non-Electric Heating	–1.3					
Mobile Air Conditioning				2.5		
Human Amenity						12.0
Human Mortality and Morbidity	5.8	11.4		9.4		37.4
Migration	0.5	0.6				1.0
Cyclones	0.8	0.2				0.3
Leisure Activities	1.7					
Water Supply					–3.0	
Availability	7.0	15.6		11.4		
Pollution				32.6		
Urban Infrastructure	0.1					
Air Pollution	3.5	7.3		27.2		
Totals						
Billions	*61.1*	*69.5*	*55.5*	*139.2*	*56.0*	*74.2*
% of GDP	*1.1*	*1.3*	*1.0*	*2.5*	*1.0*	*1.5*

* Estimates for North America.

Source: IPCC (1996a); Mendelsohn et al. (2000).

provides two reasons for the difference in sectoral estimates. Firstly, the uncertainties underlying the scientific consequences of the impacts of climate change drive different assumptions about the economic impacts. Also, the quantitative impacts may be the same but the authors can value them differently in economic terms, such variation can occur for instance in the value of a statistical life for making mortality impact estimates, an estimate that is controversial in economics and can vary widely. While this table demonstrates that there has been no shortage of estimates for the United States such comprehensive treatments are much rarer for developing regions such as SEA. One of the objectives of this book is to attempt to address this situation.

3.3 The likely impacts of climate change on South East Asia

An assessment of the magnitude of the overall impacts of climate change upon SEA is yet to be determined. However, Watson, Zinyowera and Moss (1998) went some of the way by gathering a comprehensive group of sectoral impact estimates made for tropical Asia without attempting to combine them for an overall economic impact. Gaining an indication of the vulnerability of a region of the world to climate change by making climate change impact assessments is becoming an increasingly important exercise (Olmos 2001). It allows policy makers some insight into what challenges may lie ahead with respect to options for adapting to climate change effects, an issue further examined in Chapter 7 of this book. It is also relevant for policy makers deciding emission mitigation levels, as a high vulnerability would increase the priority to reduce domestic emissions and also encourage others to as well, in order to avoid or delay climate change effects. This is all relevant for SEA, which is potentially one of the most vulnerable regions of the planet (Amadore et al. 1996). Vulnerability to climate change is defined by Reilly (1996) as the potential for negative consequences that are difficult to ameliorate through adaptive measures given the range of possible climate changes that might reasonably occur. Issues of vulnerability to climate change have been examined illustratively by authors such as Schimmelpfennig and Yohe (1999) and Yohe (2000). It can be seen from Table 3.2 that climate change can

Table 3.2 Types of Impacts Resulting from Climate Change

Climate Change Impacts					
Damage to Property	**Ecosystems Loss**	**Primary Sector Damage**	**Other Sector Damage**	**Human Well Being**	**Risk of Disaster**
– *Protection Costs*	– *Wetland Loss*	– *Agriculture*	– *Energy*	– *Human Amenity*	– *Storm/Flood*
	– *Sea Level Rise*	– *Fishery*	– *Transport*	– *Morbidity*	– *Cyclone*
– *Dryland Loss*	– *Other Ecosystems*	– *Forestry*	– *Water*	– *Migration*	– *Drought*
			– *Construction*	– *Air Pollution*	
			– *Tourism*		

Source: Fankhauser (1995b).

impact upon the economy in many different ways throughout several sectors.

3.4 Sectoral climate change impact estimates for South East Asia

This section begins by estimating the impact for each sector that would occur given $2 \times CO_2$ conditions in SEA. These impacts are given as a percentage of GDP. After they are calculated they are summed together to arrive at a total impact figure for SEA. The next step is to construct an impact function for the SEADICE model based on these results, which will be used in Chapter 4. The aggregate impact function will be discussed later in this chapter. Impacts for several different climate sensitive sectors for SEA are estimated. They are as follows:

1) Sea level rise.
2) Agriculture.
3) Health.
4) Human settlements and ecosystems.
5) Natural disasters.
6) Other vulnerable sectors.

3.4.1 Sea level rise

The estimation of future impact from SLR has been one of the most studied sectors in climate change impact assessments (Fankhauser 1995b). The importance of these types of studies is highlighted by the fact that populations are already concentrated near the coast and coastal populations are growing at twice the global average (Bijlsma et al. 1996). Many attempts have been made to measure the impact resulting from various levels of SLR (McLean and Mimura 1993; Turner, Adger and Doktor 1995; Fankhauser 1995b; Leatherman 1996). Most studies have been on either developed nations or those developing nations perceived to be most at risk, such as the small island states or Bangladesh. At this time the author is unaware of any studies that have focused on the area encompassing SEA, however, studies have been done on a global basis (Delft Hydraulics 1993).

The potential impacts of SLR can take many forms including these identified by IPCC (1994):

1. Inundation and displacement of lowlands and wetlands.
2. Coastal erosion.
3. Intensification of coastal storm flooding.
4. Increase in salinity of estuaries, salt water intrusion into freshwater aquifers, and degradation of water quality.
5. Tidal changes in rivers and bays.
6. Change of sediment deposition patterns.
7. Reduced light reaching sea floor.

This highlights the diversity of the possible effects of SLR and gives some indication of the difficulty of making estimates of any economic impacts.

3.4.1.1 Sea level rise studies for South East Asia

The coastal areas of SEA are very important not only in global terms but also for the environment, economy and society of the region. SEA comprises 29% of the world's coastlines due to the abundance of islands in its relatively small area. The coasts are home to mega-cities such as Bangkok, vast mangrove forests, busy trade ports, tourist destinations and fisheries. Other vulnerabilities are also present, 60% of the animal protein consumed in the region is derived from the sea (Yong 1989). SEA is not only prone to SLR resulting from climate change but also other anthropogenic causes of SLR. Cities such as Bangkok and Jakarta have experienced significant subsidence problems resulting from groundwater extraction. In Bangkok, the land has subsided 20mm/yr since 1960, this makes these large urban areas much more susceptible to SLR resulting from climate change (Watson, Zinyowera and Moss 1998). Recordings of SLR are being made in the region, for example SLR of 1.9mm/yr has been recorded at Hondau in North Vietnam, which is in line with globally recorded sea level changes. All of these factors indicate that SLR has the potential to be a significant impact of climate change.

The list of studies that have made estimates for some of the impacts of SLR on SEA is quite extensive, however, none give the overall sectoral impact figures for the region the SEADICE model requires. Asian Development Bank (1994) calculated the costs of a

60cm rise in sea level for Indonesia as totalling $US11.3 billion annually in sacrificed socioeconomic activity, made up of:

- 800,000 Ha of irrigated rice fields;
- 20% of 5.5 million Ha of marshlands currently used for tidal rice fields;
- 100% of the 300,000 Ha of coastal fish ponds; and
- 25% of the 4 million Ha of mangrove forest.

Recent losses of mangrove forests have been significant in SEA, with 60% losses in Philippines and 55% loss in Thailand over the last 25 years, and 37% in Vietnam and 12% in Malaysia between 1980 and 1990 (Watson, Zinyowera and Moss 1998). For Malaysia the coastline comprises 51% sandy, 42% mangrove, 6% rocky and 1% man made, of its 4,800km length. Therefore, more than 90% of the coastline is highly erodable and highly vulnerable to any increase in SLR or storm activity. A 1m SLR would cause a landward retreat of up to 2.5km in parts of Malaysia, would threaten 4200 Ha of productive agricultural land and 0.63% of Malaysia's paddy rice area (Iglesias, Erda and Rosenzweig 1996). At Las Pinas in Manila Bay it has been estimated that it would cost $US0.6 million per km to build a 3m by 1m above and a 1.5m by 3m wall below sea level (Perez et al. 1994). Using detailed maps Perez et al. (1994) found that a 2m SLR would inundate areas up to 3km inland in Manila Bay. It was found that whole towns and sections of highway were also vulnerable. Nicholls, Mimura and Topping (1995) estimate that a 1m SLR would displace over 2 million people in Indonesia, with subsequent land losses of 34,000km^2, losses of 7,000km^2 would be experienced in Malaysia and 20,000–25,000km^2 for the Mekong and Red River deltas. Nicholls, Hoozemans and Marchand (1999) estimate that the average annual number of people flooded in SEA could rise from 1.7 million in 1990 to 43 million in the 2080s as a result of SLR. Other studies to examine various aspects of the impacts of SLR in SEA include Teh and Voon (1992); McLean and Mimura (1993); Chou (1994); Midun and Lee (1995); Erda et al. (1996) and Milliman and Haq (1996).

3.4.1.2 An impact estimate for sea level rise for South East Asia

The many impact estimates just cited reveal the range of potential impacts and the serious potential for damage in the region resulting from SLR. Unfortunately at this stage the data these estimates provide is not suitable to derive a monetary estimate directly from them. In this book the method used by Nordhaus and Boyer for this sector is employed. Nordhaus and Boyer (2000) use a Coastal Vulnerability Index (CVI) to determine the relative danger faced by each country from SLR. The index is based on United States' SLR damage estimates. Each country is then compared to the United States according to coast length and total land area. Nordhaus and Boyer estimate a WTP of 0.1% of income for preventing SLR as a result of a 2.5°C warming. Using the same CVI method it was estimated that SEA is 11.29 times more vulnerable to SLR than the United States. While this is a very simple way to measure coastal vulnerability, when the special characteristics of SEA's coast are considered and the fact that SEA has been identified as one of the most vulnerable regions to coastal flooding (Nicholls, Mimura and Topping 1995) the final impact figure for the sector is not unrealistic. It was then a matter of multiplying the vulnerability factor of 11.29 by the 0.1% of GDP WTP. This resulted in an estimate of 1.13% negative impact on GDP for SEA as a consequence of SLR associated with 2.5°C warming. Given the results of the earlier stated SLR studies for the region which described some quite serious specific sectoral impacts this figure seems reasonable.

3.4.2 Agriculture

Along with SLR the agricultural sector is the other most studied area of climate change impact (Mendelsohn, Nordhaus and Shaw 1996; Frisvold and Kuhn 1999). Despite the impressive rate of industrialisation and economic growth in SEA in recent decades, agriculture remains the sector employing the most people in all of the larger countries in the region of SEA. Thus it is important in relation to climate change as its productivity is significantly affected by climate (Bazzaz and Sombroek 1996). The agricultural sector in the developing world is characterised by several factors such as; agriculture constituting a relatively larger part of the total economy, higher dependence on natural factors for production (less reliance on machinery), a significant subsistence sector and generally weaker economic support in the form of government assistance and institutional support (Luo and Lin 1999). Agriculture in SEA is largely based upon species native to the region and is dominated by rice which makes up over 90% of total agricultural production in the region. The climate change effects on agriculture are the most analysed of any climate sensitive sector.

The agriculture sector in SEA is undergoing many changes at the moment. In much of SEA rice consumption per capita is already declining, this reduces demand pressures on basic food staples while at the same time increasing demand for maize and coarse grains for stock feed (Rosegrant and Ringler 1997). In Asia crop area will increase by less than 2% by 2020 (Rosegrant and Ringler 1997), and nearly 80% of the potentially arable land is already under cultivation (World Resources Institute 1997). The total area used for growing 13 major food crops has decreased slightly since the late 1960s from about 800 million Ha, therefore increases in production have come primarily from improvements in yield. This has occurred because it has been more profitable to follow intensive agriculture than develop new land, especially where yields are already low (Budyko 1996). In recent decades global agricultural output has reached all time high levels and rates of growth (Food and Agriculture Organization of the United Nations 1996). The bulk of this increase in productivity can be attributed to the use of fertiliser and pesticides, advances in capital intensive farm management, irrigation expansion and development and selective breeding of high yielding and pest resistant crop varieties. Parikh (1994) examined the population supporting capacity for agriculture production in developing countries and estimated that a population of 22.1 billion could be supported using available land. This theoretically implies that although land supply is limited in regions such as SEA the productive capacity of the land may not be a limiting factor through which climate change would break carrying capacity.

A range of estimates, for the impact of climate change on agriculture for the countries of SEA can be seen in Table 3.3. It can be seen from this table that results vary quite widely with both positive and negative impacts forecast. This is partly due to the inherent uncertainties involved with this type of analysis and also the range of techniques and models used. Matthews et al. (1997) finds that there will be an average 3.8% decline in rice production in Asia.[16] The qualification is provided that given the time scale involved it is highly likely that management practices will adapt in response to a slowly changing climate. In Fischer et al. (1996) it was found that simulations with low level adaptation techniques compensated for climate change effects incompletely, especially in developing countries. Static climate change scenarios were run on many regions using different adaptation assumptions and Global Circulation Model (GCM) scenarios. Their main conclusion is that economic adaptation can largely compensate for moderate yield changes, however no cost estimates were made for the adaptation methods chosen.

Authors such as Watson, Zinyowera and Moss (1998) claim that the diversity of rice yield impact results for the region (displayed here in Table 3.3) demonstrates that any single estimate for the region would mean very little. While significant uncertainties exist in the estimation of climate change impacts for agriculture in SEA a single estimate for impact across the region is justified in this book on the following grounds. The region has a more homogenous climate than others, for example the countries of SEA are largely tropical monsoon climates with similar weather patterns. If an overall estimate were made for a region such as Australasia it would be less reliable simply for the reason that the region spans several

Table 3.3 Results from Impact Studies on Agriculture in South East Asia

Study	Country	Crop	Yield Impact (%)
Tongyai (1994)	Thailand	Rice	–17 to +6
Amien et al. (1996)	Indonesia	Rice	–1
Escano and Buendia (1994)	Philippines	Rice	–21 to +12
Parry et al. (1992)	Indonesia	Rice	–4
		Soybean	–10 to +10
		Maize	–25 to –65
	Malaysia	Rice	–12 to –22
		Maize	–10 to –20
		Rubber	–15
	Thailand	Rice	5 to 8
Matthews et al. (1997)	Indonesia	Rice	–6 to +22
	Malaysia	Rice	+21 to +26
	Myanmar	Rice	–9 to +30
	Philippines	Rice	–2 to +12
	Thailand	Rice	–20 to –34
Qureshi and Hobbie (1994)	Indonesia	Soybean	–20
		Maize	–40
		Rice	–2.5

climatic zones. In any case it can be argued that aggregation should not be an overarching issue, given that even within a small area such as a province, yield results can be diverse as a result of climate. Therefore, even though results may be diverse within a sample group it should not prevent the use of an overall estimate.

3.4.2.1 An impact estimate for agriculture in South East Asia

For this sector the results provided by Darwin et al. (1995) are used. They estimated the impact of $2\times CO_2$ climate change on SEA agriculture using the Future Agricultural Resources Model (FARM) which ranged from 0.2% to 1.3% of GDP. These results were used because of the coverage and depth of the study, the fact that it is a widely respected study and the fact that it provides monetary impact estimates specifically for SEA. The FARM model is a comparative statics, multiregional, general equilibrium model, covering eight world regions and 13 commodities. Using their results, their middle $2\times CO_2$ impact estimate of a –0.9% effect on GDP is the figure used here for the impact of climate change for agriculture in SEA.

3.4.3 Impacts of climate change on health

Climate change can impact upon health in several ways. Many forms of disease are climate related, either by the climate enabling the formation of a disease or by climate supporting the lifeforms that carry the disease (Kovats et al. 2000). For the tropical climates of SEA diseases such as malaria and dengue fever can be influenced by climate. The health impacts of climate change for SEA are estimated in this section.

Currently no comprehensive estimates exist for the impacts of climate change on health although it is an area that is being given special attention by academics in the region (Woodward, Hales and Weinstein 1998). Using data from Murray and Lopez (1996) and Method A used by Nordhaus and Boyer (2000, Chapter 4) for health it was found that the economic gains from the reduction in years of life lost (YLL) from climate related disease in SEA is forecast to amount to 1.15% of GDP by 2020. The details with respect to the estimated impacts for each disease type can be seen in Table 3.4. The data from Murray and Lopez (1996) provides estimates of the extent of improvements in health care in terms of the saving of the YLL from certain diseases. In this case the reduction in YLL for

Table 3.4 Effects of Climate Change on Health in South East Asia

Baseline Impact of Global Warming on Climate Related Diseases				
	Malaria	**Tropical Cluster**	**Dengue**	**Total**
Reduction in years of life lost by 2020 ('000)	2,060	474	211	2,745
Adjusted for per capita GDP×2 (% of GDP)	0.86	0.20	0.09	1.15
% impact on GDP if climate change results in 50% reduction in health gains	0.43	0.10	0.04	0.58

malaria, tropical cluster and dengue fever were obtained. The next step is to convert these advances in health outcomes into monetary values. This is done by following the Nordhaus and Boyer (2000) assumption that a YLL is worth two years of per capita income. In order to estimate the impact of climate change on health using this data the assumption is made, just as Nordhaus and Boyer (2000) did, that climate change causes a percentage loss of the business as usual economic gains in health. For the baseline scenario just mentioned it is assumed that 50% of gains are lost as a result of climate change, resulting in a cost to SEA of 0.58% of GDP. The Murray and Lopez (1996) figures are projected for the year 2020, which is some time before $2\times CO_2$ conditions are expected. They will be used in this book for, as to the author's knowledge, they are the best approximation that is available at this time. It is also a reasonable approximation since, if anything, they should tend to understate the actual impact due to the approximation date being before the expected date of $2\times CO_2$ conditions.

3.4.4 Human settlements and ecosystems

3.4.4.1 Human settlement

The possibility of future SLR as a consequence of climate change results in the potential for the significant need for human resettlement plans of action and the subsequent costs associated with them. Currently estimates of this type have been very limited.

Experience has shown that within SEA non-climate change related resettlement efforts by government have generally been expensive and had limited success (Chan 1995). However, if SLR occurs decisions will have to be made regarding protection of coastal property and/or resettlement of affected populations.

As Chan (1995) points out, physical relocation of the population under risk of inundation is an absolute last resort due to the difficulties involved including, expense, political and ethnic sensitivities and its low success rate as evidenced from past flood relocation schemes of the Malaysian Government. It was estimated by Delft Hydraulics (1993) and IPCC (1994) that 7.8 million people in SEA are at risk as a result of a 1m SLR if no protection measures are taken. It was also estimated that if adaptations were implemented the number of people at risk would be reduced to 800,000. In Indonesia approximately 110 million out of 179.4 million live in coastal areas. Asian Development Bank (1994) estimates that 3.3 million people will be displaced in Indonesia alone by 2070. These figures were obtained by adding the population from only the seven most vulnerable regions in Indonesia. Therefore, it could be argued that these figures are conservative given that most of the coastline of Indonesia was not considered. The calculated cost of relocating these people is $US8 billion (based on 800,000 homes at $US10,000 each).

While the previous estimates seem to have potential for significant impact they are even more highly speculative than normal climate change impact estimates. Therefore, they will not be used within the model implemented in this book. For the purposes of this book the figure used for the impact of climate change upon human settlements is derived from Nordhaus and Boyer (2000). The figure used is 0.10% of GDP. Given the estimates provided earlier in this section this figure can be considered as low.

3.4.4.2 Ecosystem loss

Lewandrowski et al. (1999), through the FARM model, estimate the economic impact of protecting ecosystem diversity through retiring land for conservation purposes. For SEA they estimate that land retirements of 5%, 10% and 15% would cause percentage reduction in GDP of 0.4%, 0.9% and 1.3% respectively. This result was the most severe for any region in the study, giving some indication of the relative vulnerability of SEA's biodiversity. Nicholls, Hoozemans and Marchand (1999) found that between 6 and 22% of the world's coastal wetlands could be lost under $2{\times}CO_2$ conditions. Boonpragob and Santisirisomboon (1996) made estimates from simulations using three GCM's under $2{\times}CO_2$ conditions, that the distribution of forest types in Thailand is sensitive to climate change. It was estimated that the percentage of total forest that is subtropical forest changes from 50% down to 20% while the percentage of total forest that is tropical forest rises from 45% to 80%. This substantial change in the structure of habitat potentially has substantial consequences for most plant and animal life in Thailand. For the purposes of this book the estimates from Lewandrowski et al. (1999) are assumed to correspond with the WTP to protect ecosystem diversity from the effects of climate change and therefore, a figure of 0.9% of GDP will be used as the estimate for impact of $2{\times}CO_2$ climate change on SEA's ecosystems.

3.4.5 Vulnerability to natural disasters

Potentially the most severe effects of climate change, certainly in terms of immediate impacts are likely to result from the potential change in frequency of extreme weather events. Evidence is emerging to support the assertion that climate change may have an effect on the frequency and/or force of natural disasters. Since 1990 the increase in climate related disasters has been three to four times greater than those for geological disasters (Bruce 1999). However, the link between the incidence of natural disasters such as cyclones, storm surges, flooding or drought from climate change has yet to be conclusively proven. Natural disasters can have powerful effects upon the economy, whether it be a short term effect from a cyclone or the long term effects of a prolonged drought. Property can be destroyed, lives lost, crops ruined, causing significant dislocations within an economy or even across a region. If the economic consequences of natural disasters are taken from a vulnerability standpoint, SEA is more vulnerable as a result of being a developing region. Since the 1960s developing countries have been 5.5 times more affected by disaster measured as a percentage of GDP in comparison to developed countries (Bruce 1999). SEA is especially vulnerable to natural disasters such as typhoons and cyclones. In terms of human life, an average 3,481 lives were lost and 5,126,462 people affected annually in SEA due to natural disasters over the period 1971–95 (IFRCRCS 1997).[17]

The preceding discussion highlights that the level of economic development can be reasonably used as a proxy for the impact of natural disasters.[18] Consequently, for the purposes of this book the assumption made by Nordhaus and Boyer (2000) that for lower middle income (LMI) countries the effect of the rate of change of natural disasters caused by $2\times CO_2$ climate change will be 1.01% of GDP is used.

3.4.6 Other vulnerable sectors

Tourism is an increasingly important sector for many countries in SEA. The cultural and recreational facilities available throughout SEA are becoming more popular for intra-country and international travellers. Tourism was the fourth highest foreign exchange earner for Malaysia in 1994 (Go and Jenkins 1997); in 1994 it made up 4.74% of Malaysian GDP and 5.3% for Thailand in 1992. The climate and tourism are significantly linked as pleasant climatic conditions are often needed for the full enjoyment of tourist attractions. For example it could be assumed that a significant increase in average rainfall in an area populated by beach resorts would have a negative effect on tourism numbers over time. Therefore any changes in the climate may have significant effects on tourism as a result of these preferences. Unfortunately at this time no estimates exist on the effects of climate change on this important industry so an impact estimate cannot be made for the SEADICE model. This would be an extremely difficult exercise as very specific elements of climate would need to be estimated, not only rainfall, but important factors such as the number of daily hours of sunshine would be the type of information that would be needed. It is significant that this sector cannot be included as it has the potential for significant positive and negative impacts for different regions of the world.

The forestry sector in SEA is another market sector of the economy that is vulnerable to climate change (Bautista 1990). Darwin et al. (1996) using the comprehensive FARM model found that changes in harvest rates and per hectare forest inventories due to climate change result in economic impacts ranging from 0.17% to 1.32% of GDP using different scenarios for the effect of climate change on the forestry sector in SEA. The middle range impact figure of 0.62% of GDP will be used here as the impact on other vulnerable sectors of 2×CO_2 climate change.

3.5 Overall economic impact of climate change on South East Asia

In this section the summation of all of the previous estimates of sectoral damage is made to arrive at an overall estimate for economic damage from 2×CO_2 climate change for SEA. From Table 3.5 the results for overall impact are given for baseline, optimistic and pessimistic scenarios. It can be seen that total impact ranges from impacts of –2.6% for the optimistic scenario to –6.3% for the pessimistic scenario. Meanwhile, the baseline impact total is –5.3%. It will be examined in the next section how these results compare to other studies.

3.5.1 Comparison with other results

No other estimate of this type has been done for the SEA region to the author's knowledge, although the issue has been the subject of serious discussion (Amadore et al. 1996). The closest comparison for the estimation of damage for the SEA region comes from Tol (1996) where a damage estimate of 8.6% of GDP was made for South and SEA for 2×CO_2 conditions. This is different from the regional classification used in this book. In Tol (1996) Vietnam and Laos

Table 3.5 Total Impact of 2×CO_2 Climate Change for South East Asia (% of GDP)

Sector	Scenario		
	Baseline	Optimistic	Pessimistic
Agriculture	–0.90	–0.20	–1.30
Coastal	–1.13	–1.13	–1.13
Health	–0.58	–0.17	–0.91
Ecosystem	–0.90	0.00	–1.30
Human Settlement	–0.10	–0.10	–0.10
Natural Disasters	–1.01	–1.01	–1.01
Other Vulnerable	–0.62	0.00	–0.62
TOTAL	*–5.3*	*–2.6*	*–6.3*

were not included and the countries of South Asia such as India and Bangladesh were included.

Given the estimate of Tol, the results presented in this book do not seem too pessimistic and are within the bounds of reasonable plausibility that estimates of this type can be expected. In terms of comparison with the results of other regions the final figure of a 5.3% impact indicates that SEA is much more vulnerable to climate change than the United States which has had impact estimates largely in the range of 1–2% of GDP. Even though substantial uncertainties exist for this type of estimation, the results obtained in this chapter do support the widely held belief that the developing world is more vulnerable to the effects of climate change than the developed world.

3.6 Conclusion

Whilst quite a few estimates have been made for the total economic impact of climate change for developed countries such as the United States, attempts for developing countries or regions are quite rare. This chapter provided the first known estimate of the total economic impact of $2 \times CO_2$ climate change conditions for SEA. The main finding was that a –5.3% impact on GDP is predicted for SEA, indicating that the region is relatively vulnerable to the impacts of climate change. Climate change impact estimates are needed to determine some of the policy options that are available for the region, in particular for adaptation policies, as these policies are dependent on estimates of the extent of climate change impacts. The policy implications of these impact estimates will be discussed later in the book. These results are also essential for the implementation of the impact function for the SEADICE model developed in the next chapter.

4
Model Forecasting the Future Scenarios for Climate Change and Economic Growth for South East Asia

4.1 Introduction

The estimates made in Chapter 3 for the aggregate economic impacts of $2\times CO_2$ climate change for SEA will be used towards the main objective of this chapter, which is to implement a dynamic optimisation climate-economic model for the region of SEA. To achieve this objective this chapter begins by reviewing the integrated assessment models of climate change literature. A discussion then follows of the reasons behind the choice of DICE (Nordhaus and Boyer 2000) as the basis for the SEADICE model. The method of optimisation of the model implemented in this chapter is also justified with respect to the choice between GAMS and Excel. The structure of the SEADICE model is then provided, after which some forecasting results of the model are presented.

4.2 Integrated assessment models of climate change

During the past decade (since 1992) a new group of applied economic models have been utilised for the climate change problem, known as Integrated Assessment Models (IAM). Previous to 1992, only two climate change IAMs existed in the literature (Nordhaus 1989, 1991; Rotmans 1990). IAMs are modelling frameworks that incorporate knowledge from more than one discipline (CIESIN 1995; Khanna and Chapman 1997; Kelly and Kolstad 1998). Climate change is a problem involving many disciplines and consequently has necessitated the use of IAMs that combine knowledge from both the scientific and economic disciplines. They have been used extensively to examine the problems associated with the economic effects of climate change and to provide economically efficient policy solutions. The main function of these models should be to allow the determination of the implications of stylised relationships between environmental and economic systems and simple but explicit specifications of value judgements. Currently IAMs of climate change need to at least include some reduced form modules, as the complexities of the problem are too great at this time. As Janssen (1996) points out, the most challenging aspect of building an IAM is getting the balance right between factors such as simplicity and complexity; stochastic and deterministic elements; aggregation and realistic outcomes; qualitative and quantitative linkages; transparency and uncertainty.

As with any type of modelling that involves economics, aspects of it will be controversial. Rotmans and Dowlatabadi (1996), reveal the main disadvantages of IAMs which include:

1. Too complex a structure.
2. High level of aggregation.
3. Explication of counter intuitive results.
4. Insufficient treatment of uncertainty.
5. Absence of stochastic behavior.
6. Limited verification and validation.
7. Inadequacy of knowledge and methodology.

Whereas, some of the advantages of IAMs are described by Janssen (1996):

1. The use of system interactions and feedback mechanisms.
2. Can be used and replicated for the purposes of experimentation more easily due their simplicity.
3. The weaknesses in some areas of scientific knowledge can be identified.
4. Improved communication between scientists and modellers is facilitated.

These strength and weaknesses indicate that IAMs share many characteristics of economic modelling in general and that as always both strengths and weaknesses must be acknowledged in any study that uses an IAM. An excellent critical overview of IAMs of climate change can be found in Khanna and Chapman (1997).

In this section a survey of the relevant literature is provided. A particular strand of the IAM group; that of *policy optimisation models* is the focus of the review. Policy optimisation models calculate the optimal policy control variables, given a formulated policy goal. The most well known policy optimisation models are DICE (Nordhaus 1994a), MERGE (Manne, Mendelsohn and Richels 1995), CETA (Peck and Teisberg 1995) and FUND (Tol 1996).

The CETA model is a derivative of the Global 2100 model originally developed by Manne and Richels (1992). The CETA model was developed by Peck and Teisberg (1993; 1995) and is a set of models consisting of a single world region that includes component models for the carbon cycle, climate change and impacts. Illustrative damage functions are defined that represent climate change damage at any time as an increasing function of the change in global average temperature. These damage functions are calibrated to the Nordhaus upper estimate of 2% GDP loss for $2\times CO_2$ conditions. In another incarnation of CETA Peck and Teisberg (1995) consider the advantages and disadvantages of international cooperation using a two-region version of CETA. The model predicts that a reduction of worldwide emissions by one-third over what they would be in the baseline case would produce benefits of about $US1.2 trillion worldwide if the developing countries were not participating in reduction policies, or about $US1.5 trillion if they were.

MERGE (also based on Global 2100) is a dynamic general equilibrium model with five world regions and a single consumer in each region who makes both savings and consumption decisions (Manne, Mendelsohn and Richels 1995). Impact functions are defined for both market and non-market components, where market impacts are a quadratic function of temperature change, and also calibrated to the estimates of Nordhaus. The non-market impacts are modeled as a worldwide public good, where each region has a WTP to avoid a specified temperature change represented by an S-shaped function of regional income. A distinctive characteristic of the model is that it includes international trade in oil, natural gas and energy-intensive basic materials.

The DICE model, developed by Nordhaus (1994a) is a dynamic integrated model of climate change in which a single global producer-consumer makes choices between current consumption, investing in productive capital, and reducing emissions to slow climate change. This model is used as the basis of the SEADICE model and a thorough review will be provided later in this chapter.

The Framework for Uncertainty, Negotiation and Distribution (FUND) model was originally created to study the role of international capital transfers in climate policy (Tol 1995). It then evolved into a model that examined the impacts of climate change in a dynamic context. FUND, like the other models reviewed here links simple models of population, technology, economics, emissions, atmospheric chemistry, climate, sea level and impacts. An important aspect of the FUND model is that it estimates monetary impacts due to both the rate and level of climate change.

4.2.1 Major modelling issues

Although to a substantial degree many aspects of this young branch of economic modelling have converged such as the use of baselines and certain climate submodels, there are currently many important issues that are still providing particular difficulties. Some of these major modelling issues will be discussed to provide further background to the later presentation of the SEADICE model.

4.2.1.1 Data limitations

A major limitation for IAMs is that there are substantial data limitations with regard to the economic impacts of climate change which are needed to calibrate the impact functions of these models. These limitations manifest themselves in two forms; firstly the lack of suitability of many scientific impact studies for the conversion of their data to monetary impacts. Secondly, many studies still rely on scientific data for the United States and other OECD countries and then extrapolate their results to the rest of the world. There are numerous examples of the sometimes ad hoc treatment of the estimation of impacts for developing nations. The MERGE model simply assumes that damage in developing nations is twice that of developed nations, and the PAGE model uses a multiplicative factor across all impact sectors for each non-European Community (EC) region. However, over time the number of scientific studies on developing regions is steadily increasing. Therefore, it is becoming an increasingly viable option to model climate change effects for non-OECD regions. The increased availability of quality data is why the estimates that were made in Chapter 3 are now possible.

4.2.1.2 Intertemporal issues

As always, discounting is a controversial issue and in the case of climate change models it is very important due to the long time frames involved (Nordhaus and Boyer 2000; Tol 1999). Without elaborating on the fundamental differences between those that propose positive, zero or even negative discount rates, this book will simply observe the discounting options chosen in the narrow field of interest of this book, and the importance of the rates used to overall results. In general rates of between 0 and 3% have been used in this type of modelling.

SEA faces the same intertemporal equity issues raised by climate change as all other regions. That is, how much will today's generation value the welfare of future generations? The discount rate has been a controversial concept in economics for quite some time, and it is one that does not look likely to be resolved. This is primarily because the various points of view are distinguished on not only economic but also ethical grounds. The standard representation of the Social Rate of Time Preference (SRTP) is as follows:

$$\text{SRTP} = \rho + \theta g$$

where

ρ = the rate of pure time preference,
θ = the absolute value of the elasticity of marginal utility, and
g = the growth rate of per capita consumption/income.

The above formula represents the sum of pure time preference (impatience) and the rate of increase in the welfare derived from higher per capita incomes in the future. In other words you either care less about future consumption than present or you believe future consumers will be better off than today's. Economists are in general agreement about the form of the SRTP. However, disagreement exists regarding several factors that influence the rate of discount. These include methods of analysing uncertainty of forecasted variables, the likely rate of future per capita economic growth and the conversion of investment into consumption equivalents. There are two main approaches to the estimation of the discount rate in climate change economics; prescriptive and descriptive. Each will briefly be explained in turn.

Four main views are emphasised by those who follow the prescriptive approach as described by Arrow et al. (1996): (1) market imperfections and sub-optimal tax policy; (2) policy constraints, in particular the problems with intergenerational transfers; (3) distribution for equity, by using a low discount rate some efficiency will be lost but the gains in equity are enough for it to be justified; (4) the goal of the equalisation of the marginal utility of consumption over time. As a result of these views and conclusions the prescriptive approach generally results in the use of low discount rates for the changes in consumption of future generations.

The descriptive approach focuses on the opportunity costs of capital and is used for most climate change optimisation models such as those formulated by Peck and Teisberg (1993; 1995), Nordhaus (1994a) and Manne, Mendelsohn and Richels (1995). The descriptive approach has three main arguments: (1) that mitigation expenditures will displace other forms of investment; (2) if the rate of return of investments other than mitigation are higher than mitigation then society would be better off with the investment with the higher rate of return; (3) the appropriate social welfare function to use for intertemporal choices is revealed by society's actual choices, therefore the social discount rate (SDR) should be equivalent to current rates of return and growth rates. Critics of the descriptive approach have argued against all three of these points. There is also the classic ethical argument against any discount rates above zero because they devalue future generations.

For SEA the rate of economic growth is likely to be quite high as many of the nations in the region continue their race towards industrialisation. Therefore, it is more likely that the descriptive approach would be more useful to provide an estimate of the discount rate. Higher growth rates imply that future generation should be more able to deal with any climate change effects that may occur. Also the expected high returns of investment in these economies is more likely to crowd out mitigation and adaptation policies that are not viable because more money can be made with other projects. Using a higher discount rate such as the 3% used by Nordhaus and Boyer (2000) will go some way towards accounting for these effects.

4.3 The choice of model

The model implemented in this book is based upon one of the most famous models in the literature, Nordhaus' DICE model (Nordhaus 1994a; Nordhaus and Boyer 2000). The DICE model combined the economy and climate in a dynamic optimisation framework for the first time. The model is based upon the Ramsey model of optimal economic growth and consists of two sectors, the economic sector and the climate sector. In the economic sector only one good is produced competitively and is perfectly substitutable. The social planner allocates the good between consumption and investment to optimise intertemporal utility. The production function is Cobb-Douglas with an emission damage term included. Labour input and technological change are both assumed to be exogenous, and to have growth rates that decay exponentially. Resources are allocated from the economic sector to the climate sector by the social planner to minimise the negative effects of global warming. The climate sector consists of several equations that define the relationship between GHG emissions and economic activity. Atmospheric absorption of GHG and the ocean's role as a thermal absorber are represented by separate equations.

The problem of climate change is one that occurs over decades and centuries, 'Dynamics are therefore of the essence' (Nordhaus 1994a, p. 5). The majority of changes from global warming are gradual and occur over many years. Consequently, the speed and dynamics of the change are very important for economic growth. The estimated residence time of CO_2 in the atmosphere is over 100 years, which illustrates how long the time lags exist in many aspects of climate change analysis. The neoclassical optimal growth model that DICE and SEADICE are based upon will be used to incorporate the dynamics of economic growth over time, which is an advantage for examining climate change. The DICE model has been controversial over the last decade, the arguments for and against will be examined in Section 4.3.2.

4.3.1 The difference between DICE and SEADICE

The SEADICE model implemented in this chapter is different from the original DICE model. The main difference is that it represents a region (SEA) whereas the DICE model is global. An example of a model that has successfully adapted the DICE model from a global to a country level, to a region other than the United States is the Australian Dynamic Integrated model of Climate and the Economy (ADICE) of Islam (1994). ADICE derived results for Australia by splitting world CO_2 emissions into an Australian and a ROW component. The ROW component is exogenously determined in the model, leaving only the introduction of Australian economic data to estimate climate change impacts on that country. This same method was also used for China with the DICE-CHN model of Jia (1996). This method will also be used for the SEADICE model in this chapter. In the SEADICE model ROW emissions are exogenous and obtained originally from the results of the DICE model. This method is explained in more detail in Section 4.4.3 within the description of the climate module of the SEADICE model.

The other major difference between the DICE model and the SEADICE model is that many of the parameter and variable values for the model have been specifically chosen to represent SEA. The impact estimates made in Chapter 3 are an example as well as many other parameters and variables such as population, capital, etc. The full list of parameter and variable values can be seen in Table 4.A.2 in Appendix 4.A.

4.3.2 Arguments in favour of the DICE model

This section presents the arguments regarding why DICE was chosen as the framework for the modelling of the economics of climate change in this chapter. The strengths of the DICE type model framework compared to others can be explained as follows:

1. Ease of use – this model is transparent, therefore more portable, understandable and transferable, also more likely to be used to broaden the understanding of some of the relationships between climate change and economic damage to a wider audience including policy makers. As Krugman highlights, 'useful fictions' are those modelling devices that make use of simplified laws to cut through the complexities of the world, 'models are metaphors ... we should use them, not the other way around' (1996). In Global Environment Facility (2000) it was found that 'many countries reported frustration at the lack of materials and software for carrying out technical studies, in particular those

related to projections and modelling. In some cases, the cost of relevant software is prohibitive ...' (p. 52). With the SEADICE model implemented in this chapter this book is promoting the use of a relatively easy to implement model. This type of modelling while possessing some weaknesses as a result of its simplicity, does have the strength that it is easy to implement, and therefore more likely to be further developed by a wider range of researchers.
2. It shares the advantages of other IAMs already mentioned in Section 4.2. These advantages were identified by Janssen (1996) as: the use of system interactions and feedback mechanisms; the possibility of replication for the purposes of experimentation made easier due to their simplicity; the possibility of identifying weaknesses in some areas of scientific knowledge and the facilitation of improved communication between scientists and economic modellers.
3. The model results are economically optimal. The best combination of resources possible, given the economic structure of the model, are found to arrive at a specific outcome. Economic criteria, specifically those related to the costs and benefits of a certain policy scenario are widely understood by policy makers.
4. Another advantage is that the model is dynamic. Therefore, the paths and behaviour over time of many important economic and environmental variables related to climate change can be seen. Being able to see how these variables change over time gives greater insight into the relationships of climate change economics.

These arguments present the case of why using a DICE type model to represent the economic effects of climate change is acceptable and consequently why it is used in this book.

4.3.3 Arguments against the DICE model

Since the DICE model came to prominence in 1994 it has been quite controversial and many authors have criticised aspects of the model including Broome (1992), Azar (1995), Sen (1995), Anand and Sen (1996), Costanza (1996), Janssen (1996) and Mabey et al. (1997).

A common argument against the DICE model that has been used by Janssen and others is that the model does not represent the complex behaviors and dynamics of the climate system adequately for the results to be useful. This type of argument ignores what

Nordhaus actually created with the DICE model. First of all the time frames involved in the model should be examined. DICE is decadal, therefore, values for emissions, temperature increase, etc. are all only given once every ten years. Because of this Nordhaus adapted a *simple* climate module for the DICE model, an overly complex model is simply not justified for decadal intervals.[19] Furthermore, the DICE model interacts with the economy through only one variable, temperature. Therefore, a complex model that provides dynamic estimates for temperature, precipitation, and many other climate variables is not needed. The level of complexity and dynamics that can be represented by one variable at ten year intervals is very limited. Connecting a complex climate model to the DICE model would be like trying to push a pumpkin through a garden hose. Nordhaus never represented DICE to be something that it is not, it is a typical economic model where the essence of particular relationships are examined to try to further the understanding of key elements within a complex and interrelated environment. Nordhaus and Boyer (2000) provide three reasons why simplification of the scale represented in DICE is warranted. Firstly, complex systems cannot be understood easily and also provide higher probabilities of erratic behavior. Secondly, sensitivity analysis can be undertaken more easily in a simpler model in order to determine the model's robustness compared to a larger model. Finally, personal computers are testing the limits of their performance already with DICE as it is specified now; a model that is created for widespread distribution and use for other researchers must be usable on personal computers. Therefore, simplification in this case does not render the results to be useless as posited by Janssen (1996).

DICE of course shares in common the weaknesses of the IAM modelling literature overall, which were discussed earlier in Section 4.2. While the criticisms of the DICE model have been sustained over the past few years it is argued here that it is still the best option available for the purposes of this book. The alternatives that are available at this time, to the author's knowledge, do not offer all of the advantages as outlined in the previous section.

4.3.4 The model solution process – the choice between GAMS and Excel

Another decision that had to be made was the choice of model solution process. Due to the recent creation of an Excel version of DICE by Nordhaus the option now exists to solve the model with either Excel or GAMS software.[20] In this section an extrapolation is made from Nordhaus and Boyer (2000) about the various differences between solving the model with GAMS and with Excel, clearly stating the reasons why Excel is used for the experiments in this book.

There are several reasons why Excel was used in preference to GAMS as the optimisation tool for the model:

- The authors have extensive experience in Excel and could therefore utilise the full capabilities of the model more quickly and confidently than the alternative.
- Excel offers significant opportunities to create a wide array of output from the model such as tables, graphs, and additional variables from the model output.
- Nordhaus and Boyer (2000) explain the accuracy of GAMS is superior to that of Excel for the concentrations limit case. However, since this case is not relevant to this book (as described in Section 6.1.5) this anomaly is not important.
- Excel solution time turns out to be faster compared to GAMS.
- Excel is more user friendly, where changes in parameters, etc. can be seen in an instant. GAMS requires the user to run the program again so that changes can be observed. This advantage also makes the model more accessible to end users in the countries of SEA.
- Excel is far more accessible to the developing nations that the SEADICE model attempts to represent. Therefore, there is a far greater likelihood that scientists and economists from SEA might develop the SEADICE model to better represent the region.

For the above reasons Excel was chosen over GAMS as the optimisation tool for the SEADICE model.

4.4 Model structure

Given that DICE (Nordhaus and Boyer 2000) has been chosen as the framework for the SEADICE model[21] for the reasons outlined in this chapter it can now be outlined in detail. In the following sections the SEADICE model is explained equation by equation revealing the structure and relationships of the model and the purpose of each equation.

4.4.1 Objective function

In order to solve the SEADICE model there must be an objective function so that utility can be measured and optimised. With respect to utility, the major economic assumption made is the classic 'consumption is good' argument where more consumption is preferred over less. In modelling, this is embodied in an equation where the objective is to maximise the discounted sum of the utility of per capita consumption. The function is represented as:

$$\max_{\{c(t)\}} \sum_t U[c(t), L(t)](1+\rho)^{-t}$$

where

U = the utility of society
c(t) = consumption per capita at time t
L(t) = the population level at time t, and
ρ = the pure rate of social time preference.

Discounting applies to utility not monetary values in this objective function. Therefore, it specifies a value judgement about the distribution of utility across generations. The discount rate is assumed to decline over time because of the assumption of declining impatience (Nordhaus and Boyer 2000). The rate of time preference starts at 3% per annum in 1995 and declines to 2.3% per annum in 2100 and 1.8% per annum in 2200.

The explicit form of the utility function is assumed to have the constant elasticity of substitution (CES) specification, hence;

$$U[c(t), L(t)] = \frac{L(t)\{[c(t)]^{1-\alpha} - 1\}}{(1-\alpha)} \tag{1}$$

where α = elasticity of marginal utility of per capita consumption.

In the SEADICE model it is assumed that α =1, hence the logarithmic functional form is:

$$U[c(t), L(t)] = L(t).\log c(t) \tag{2}$$

Therefore, equation (1) using the explicit functional form of equation (2) is maximised subject to the economic and climatic constraints outlined below.

4.4.2 Constraints

Output is represented in the model by the standard constant-returns-to-scale Cobb-Douglas production function inclusive of technology (A(t)), capital (K(t)) and labour (L(t)). The elasticity of output with respect to capital follows the assumption of Nordhaus (1994a) of 0.25. Output is gross with respect to depreciation of capital but net with respect to climate damages and mitigation costs as represented by the term W(t) which is the output scaling factor (discussed below).

$$Q(t) = \Omega(t)A(t)K(t)^{\gamma} L(t)^{1-\gamma} \quad (3)$$

where

γ = elasticity of output with respect to capital;
= factor share of capital = 0.35;
Q(t) = gross domestic product;
A(t) = level of technology;
L(t) = labour force;
$\Omega(t)$ = output scaling factor due to emissions controls and to damages from climate change (with $\Omega(1) = 1$); and
K(t) = capital stock (at start of period t).

To make the model more representative of SEA a specific value had to be found for total factor productivity (TFP), A(t) in the above equation. For the purposes of this model the figure used in Nordhaus and Boyer (2000) for LMI countries will be used as the proxy for SEA TFP. SEA consists of nine countries of which Nordhaus and Boyer considers one to be high income (Singapore), one to be middle income (Malaysia), one to be lower middle income (Thailand) and six to be low income (Indonesia, Philippines, Vietnam, Myanmar, Cambodia, and Laos). In terms of GDP the high, middle and lower middle income group account for 45% of the GDP of the region. Indonesia by itself accounts for 30% of GDP in the region and as it is very close to being classified as LMI it was decided that LMI would be the more appropriate category to use for SEA.

The growth rate of both the technological change and the labour inputs are assumed to decay exponentially. Hence the time paths of the growth rates of the two variables are:

$$g_A(t) = g_A(t-1)(1-\delta_A) \quad (4)$$

$$g_{pop}(t) = g_{pop}(t-1)(1-\delta_{pop}) \quad (5)$$

where
g_A = growth rate of technology;
g_{pop} = growth rate of population;
δ_A = decay rate of the growth rate of technology; and
δ_{pop} = decay rate of the growth rate of population.

The assumption that output consists of either investment (I(t)) in new capital or consumption (C(t)) is represented by the equation:

$$Q(t) = C(t) + I(t) \quad (6)$$

It follows that per capita consumption (c(t)) is represented as:

$$c(t) = C(t)/L(t) \quad (7)$$

The rate of capital accumulation (K(t)) is represented by the standard equation;

$$K(t) = (1-\delta_K)K(t-1) + I(t-1) \quad (8)$$

where
δ_K = the rate of depreciation of the capital stock.

Therefore, changes in capital are a function of additional investment and the depreciation of existing stock. Following Islam (1994), the rate of depreciation is assumed to be 8%.

4.4.3 Climate equations

The SEADICE model represents the climate economy relationship with a series of simplified representations of the climate change process. This process is taken from the emission of GHG into the atmosphere from economic activity to the eventual damage effects on economic output of the resulting climate change. The following equations are very simple compared to many models of the atmosphere, for example Rotmans (1990). As discussed earlier, this is a function of necessity, as most of the scientific representations of the atmosphere are too complicated to be coupled with an economic model where transparency of results is a desired attribute.

Emissions are assumed to be proportional to output. The equation for emission output is represented as:

$$E(t) = [1 - \mu(t)]\sigma(t)Q(t) \tag{9}$$

where

$\mu(t)$ = the fractional reduction of emissions relative to uncontrolled emissions, and

$\sigma(t)$ = the uncontrolled ratio of GHG emissions to output.

The control rate (m(t)) is determined by the optimisation of the model. The value used for s(t) is derived from Nordhaus and Boyer (2000) where each value of s(t) was taken for each SEA country and an average figure taken which is 0.27. This figure reduces over time representing a gradual reduction in uncontrolled emissions. This is represented as:

$$\sigma(t) = \sigma(t-1)/[1 + g_\sigma(t)]; \tag{10}$$

where

$g_\sigma(t)$ = the rate of decadal reduction in $\sigma(t)$.

This book follows Nordhaus and Boyer (2000) by introducing emissions from land use change into the model in the following way:

$$LU(t) = LU(0)(1 - \delta_1)^t \tag{11}$$

$$ET(t) = E(t) + LU(t) \tag{12}$$

where

LU represents emissions from land use change

δ_l = the rate of decline in land use change emissions, and

ET = total emissions including land use change.

Total emissions then interact with the atmosphere and oceans to change the concentrations of GHG in the atmosphere. The rate of the accumulation of GHG in the atmosphere is represented by:

$$M(t) = ET(t-1) + E_{ROW}(t-1) + \phi_{11}M_{AT}(t-1) - \phi_{12}M_{AT}(t-1) + \phi_{21}M_{UP}(t-1) \quad (13)$$

$$M_{UP}(t) = \phi_{22}M_{UP}(t-1) + \phi_{12}M_{AT}(t-1) - \phi_{21}M_{UP}(t-1) + \phi_{32}M_{LO}(t-1) - \phi_{23}M_{UP}(t-1) \quad (14)$$

$$M_{LO}(t) = \phi_{33}M_{LO}(t-1) - \phi_{32}M_{LO}(t-1) + \phi_{23}M_{UP}(t-1) \quad (15)$$

where
$M_i(t)$ = total mass of carbon in reservoir i at time t (GtC);
$E_{ROW}(t)$ = rest of world emissions; and
ϕ_{ij} = the transport rate from reservoir i to reservoir j per unit time.

The reservoirs are:
AT = atmosphere,
UP = all quickly mixing reservoirs (the upper level of the ocean down to 100 metres and the relevant parts of the biosphere), and
LO = deep oceans.

This specification of the carbon cycle is different from the 1994 version of the DICE model. With this specification Nordhaus and Boyer (2000) responded to the possibility that the time series data of the original DICE might understate carbon retention because an infinite deep ocean carbon sink was assumed. The result was the preceding three equations where a three-reservoir model calibrated to current scientific carbon-cycle models is used. The deep ocean reservoir for carbon is massive but limited and is fed by quickly mixing reservoirs in the atmosphere and upper ocean. All three reservoirs mix well in the short run but in the long run the upper reservoirs mix very slowly with the lower reservoir.

There is also another difference in this equation compared to the original DICE model as it follows the method used in the ADICE model to distinguish a particular country. This is done by separating the emissions of SEA and ROW using the variable E_{ROW}. In the model ROW emissions are exogenous and the values were obtained originally from the DICE model. SEA emissions are determined by the model and are subtracted from the ROW emissions to avoid double counting. By making this split the SEADICE model is transformed into a game theoretic model. In this case it is a two person, non-cooperative game where the emissions from the ROW are given and SEA must optimise its strategy subject to the actions of the ROW.

Another equation that is derived from climate models is that for the measurement of radiative forcing. Radiative forcing is influenced by the GHG concentrations calculated in the previous equation. This equation is not controversial and therefore is a standard type representation portrayed as:

$$F(t) = 4.1\frac{\log\left[\frac{M(t)}{735}\right]}{\log(2)} + O(t) \tag{16}$$

$$\begin{aligned} O(t) &= -0.1965 + 0.13465t \qquad t < 11 \\ &= 1.15 \qquad\qquad\qquad\quad\;\; t > 10 \end{aligned} \tag{17}$$

where
$F(t)$ = the increase in surface warming in watts per m^2 which is a function of the accumulation of GHG in the atmosphere, and
$O(t)$ = a representation of non-CO_2 GHG.

The final climate equation is a representation of the mean temperature change at surface level, which is influenced by the radiative forcing and lags in the system between thermal layers of the ocean and atmosphere. Following Nordhaus and Boyer (2000) these equations are based on the model of Schneider and Thompson (1981).

$$\begin{aligned} T(t) &= T(t-1) + \left(\frac{1}{R_2}\right)\{F(t) - \lambda T(t-1) - \frac{R_2}{\tau_{12}}[T(t-1) - \\ &\quad T^*(t-1)]\} \\ T^*(t) &= T^*(t-1) + \left(\frac{1}{R_2}\right)\left\{\left(\frac{R_2}{\tau_{12}}\right)[T(t-1) - T^*(t-1)]\right\} \end{aligned} \tag{18}$$

where
R_1 = the thermal capacity of the upper stratum
R_2 = the thermal capacity of the deep ocean

T^* = the deviation of the deep ocean temperature from pre-industrial levels
τ_{12} = the transfer rate from the upper layer to the lower layer, and
λ = a feedback parameter.

The change in temperature derived from the previous equation is assumed to impact upon the economic part of the model; the equation that provides this link is the damage (or impact) equation. The damage equation is one of the most important of the model. This equation represents the economic damage resulting from climate change. Damage is determined by the level of GDP (Q(t)) and the temperature (T(t)) where the parameters b_1 and b_2 determine the shape of the damage curve with respect to temperature.

$$D(t) = Q(t)b_1T(t)^{b_2} \tag{19}$$

Temperature was used originally by Nordhaus (1994a) as a proxy for overall climate change. Authors such as Toth (1995) have suggested this may be a mistake as it may have taken the research community's focus from potentially dangerous changes in climate apart from temperature. However, even now, without the provision of a detailed climate model, temperature remains the best option available for dynamic optimisation models of the DICE type. This book also follows the DICE model by treating the damage function as quadratic, that is, the parameter b_2 equals 2. At this stage this value can only be based on a best guess range of between 1 and 2 resulting from the work of Cline (1992) who estimated a damage function power of 1.3 and Nordhaus (1994b) where the median result of an expert panel predicted a value of 1.5. However, others have used a damage function with a power as high as 3 (Peck and Teisberg 1995).

Other forms of damage function have been used in other models. One of the first was Cline (1992) who, like Nordhaus, assumes damage will be non-linear in nature.

$$d_t = d_0\left[\frac{T_t^u}{n}\right]^\gamma$$

where
d_t = damage as a fraction of world GDP

T = temperature at time t, and
n = 2.5°C degrees.

The damage function used by Fankhauser is of the following form,

$$D_t = k_t \left[\frac{T_t^u}{n} \right]^{\gamma} (1+\theta)^{t^* - \bar{t}}$$

where
k_t = market and non-market based damage
t^* = the time when CO_2 concentration doubles, and
$\bar{t}$ = the time when $D_t = k_t$.

Fankhauser criticised both Cline and Nordhaus because their estimates do not adequately deal with social non-market costs, which Fankhauser argues are a significant portion of total damage. All damage functions of this type are controversial as a result of their highly stylised representation. While a large amount of uncertainty exists about how well they might represent aggregate climate change impacts, at the moment they are the best available approximation for these types of models.

The total cost equation represents the cost of mitigating GHGs where the emission control rate ($\mu(t)$) is determined by the model optimisation. The parameters θ_1 and θ_2 determine the shape of the mitigation cost curve.

$$TC(t) = Q(t)\theta_1 \mu(t)^{\theta_2} \tag{20}$$

The Omega equation represents the ratio of mitigation costs to climate damage. The value of omega is included in the initial production function as the link between emissions, climate change and the economy. Omega (and therefore the economy) is negatively related to the level of climate damages and positively related to the level of mitigation costs. Put more simply, the Omega equation enables economic production to decrease if mitigation costs and/or damage are higher, and *vice versa*.

$$\Omega(t) = \frac{1 - \theta_1 \mu(t)^{\theta_2}}{1 + b_1 T(t)^{b_2}} \tag{21}$$

The carbon tax derived in Nordhaus and Boyer (2000) is calculated using the formula:

$$Ctax = -1000 * ee.m(t)/kk.m(t) \tag{22}$$

where
ee.m(t) and kk.m(t) are the shadow prices of emissions and capital.

Equations (1) to (22) described above represent the SEADICE model. The entire model, as well as a list of the major variables and starting values for the parameters of the model can be found in the appendix to this chapter.

4.5 Model results

Given that the SEADICE model has now been formulated, the next step is to implement the model and obtain results. The SEADICE model was run for five different sets of parameters and the results are presented in this section. The forecasts for all major economic and environmental variables in the SEADICE model are presented and analysed.

4.5.1 SEADICE results for five model runs

According to Islam (2001) the results of models based on the ADICE framework, and therefore SEADICE, should be interpreted in the following spatial, structural and policy framework:

1. Global warming depends on the GHG emissions of SEA and the ROW. SEA emissions depend on the economic and technical characteristics of the SEA regional economy, while the ROW emissions are exogenous in SEADICE (adopted from DICE).
2. SEADICE determines the global warming and other economic effects of SEA GHG emissions.
3. SEADICE compares the benefits and costs of SEA GHG policies to suggest the optimum policies given that the ROW undertakes an optimum policy action. SEADICE also determines the SEA unilateral optimum GHG emissions and policies assuming ROW does follow an optimum policy (ROW optimum policy trajectory is given in SEADICE).

SEADICE was solved for five sets of parameters. In the following sections, the results of five different model runs are reported. The results for each model run spans ten periods, one period being equivalent to a ten-year duration. The differences in parameter values between the five models are shown in Table 4.1. It can be seen from this table that Model Run 1 (referred to here as the Base model run) represents a situation where GHG emissions are under no control and therefore represents a baseline scenario. Model Run 2 (referred to here as the Optimal model run) is a scenario where the optimal GHG emission rate is produced. Model Run 3 (referred to here as the Technological Breakthrough model run) represents a scenario of no GHG emissions and consequently provides the forecast for the situation where no climate change impacts occur, and is therefore equivalent to the theoretical case of a massive technological breakthrough that results in no climate change effects on the economy. Model Runs 4 (referred to here as the Zero Discount Rate model run) and 5 (referred to here as the Higher Decline Rate of Uncontrolled Emissions (HDRUE) model run) present the same scenario apart from changes in the discount rate and the growth rate of uncontrolled emissions. These model runs will provide forecasts for many major economic and environmental variables for SEA as well as indicating the sensitivity of the model to important parameters.

All five parameter specifications were optimised using an Excel version of the SEADICE model. All model runs optimised successfully and within the parameters set forth in Nordhaus and Boyer

Table 4.1 Details of Model Runs

Model Runs	Emission Control Rate	Discount Rate	Growth Rate of Ratio of Uncontrolled Emissions
1. Base	No control	3.0%	–0.1168
2. Optimal	Control	3.0%	–0.1168
3. Technological Breakthrough	No GHG emissions in the model (pure economic model)		
4. Zero Discount Rate	Control	0.0%	–0.1168
5. Higher Decline Rate of Uncontrolled Emissions	Control	3.0%	–0.2168

(2000) for this type of application. In the appendices to this chapter the numerical results are reported in Table 4.B.1 (climate and environmental variables), Table 4.B.2 (economic variables) and Table 4.B.3 (different scenarios). In the following sections the results are described using Figures 4.1–4.16 as support.

4.5.1.1 Results for climate and environmental variables

In this section the results from SEADICE for the climate and environmental variables are presented. Although the parameter specification of SEADICE is based on the benchmark of 2×CO_2 assumptions, SEADICE has the potential to make projections of climate variables far in the future beyond this benchmark. However, the vast majority of studies present only 2×CO_2 results, which is the benchmark present here. The Technological Breakthrough model run assumes that there is no greenhouse effect and the climate and environmental variables do not effect the economic part of the SEADICE model. Therefore it does not appear on any of the environmental results presented here.

Climate damage as a percentage of net output is invariant to the changes of parameters made in this experiment (see Figure 4.1). This is because damage is dependent upon global temperature changes that are not sensitive to changes in policy from SEA. The reason for

Figure 4.1 South East Asia Climate Change Damage as a % of GDP

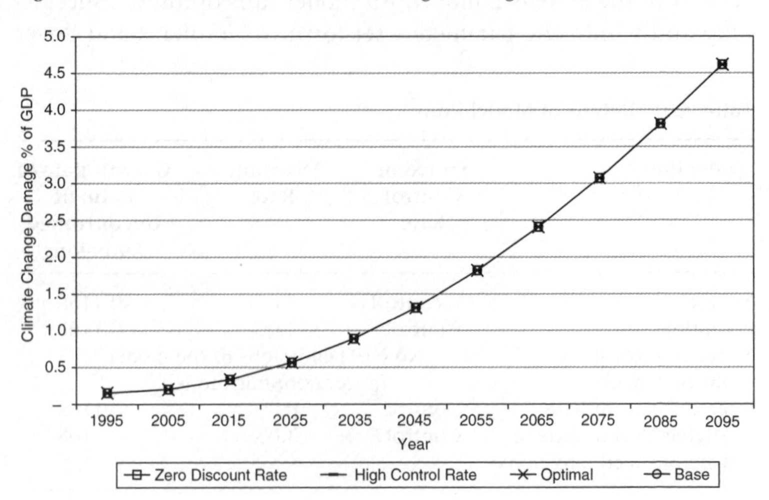

this is that SEA emissions make up only 2% of global GHG emissions. Therefore, even order of magnitude changes in SEA emissions will have little effect on overall global emission levels. This is a demonstration that future changes in temperature from climate change are largely out of SEA's hands. It is the rest of the world that will largely determine the extent to which climate will change. This is another reason why it is important that the governments of SEA prioritise adaptation strategies in order to cope with the future economic effects of climate change. From Figure 4.1 below it can be seen that $2 \times CO_2$ damages (in 2095) from climate change total almost 5% of GDP. This can also be viewed as the maximum potential benefit possible from the implementation of adaptation strategies. In other words if it is known how much damage climate change can do, then it is known how much can be spent to prevent that damage without incurring net costs.

Figure 4.2 below demonstrates the sensitivity of total SEA CO_2 emissions to the changes in parameters in the five model runs. It is apparent that an increase in the rate of decline of uncontrolled emissions resulting from the HDRUE model run significantly reduces SEA carbon emissions. This happens because the HDRUE model run causes the percentage of emissions that are controllable

Figure 4.2 Total South East Asia Carbon Dioxide Emissions

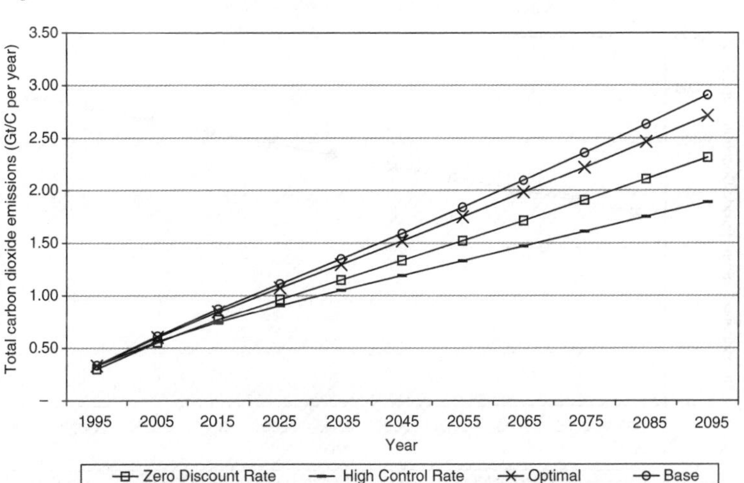

to increase over time and therefore there is more scope for emission reductions and hence total emissions are lower compared to other model runs. The paths of SEA GHG emissions for most of the model runs vary markedly. The Base, Optimal and Zero Discount Rate model runs all show higher growth in GHG emissions suggesting that they are more likely to be unsustainable in a long term policy sense. The HDRUE model run is the only model run that shows a significantly lower emissions and therefore the highest probability to be ecologically sustainable. Overall the results for emissions indicate that the model is working logically. For example, the Zero Discount Rate model run gives a lower emission profile, consistent with the current generation reducing emissions as a result of concern for the welfare of future generations.

Figure 4.3 supports the argument made in the earlier paragraph, describing Figure 4.1 in that CO_2 concentrations, which are global variables in the model are not sensitive to changes in any of the parameters changed in all of the model runs. The effect of a SEA policy of no mitigation is evident in the CO_2 concentration results. The results show that a CO_2 concentration of optimal paths between GHG mitigation policy and no mitigation policy alters the concentration of GHG in the atmosphere by a very small amount. This

Figure 4.3 Global Atmospheric Carbon Dioxide Concentrations

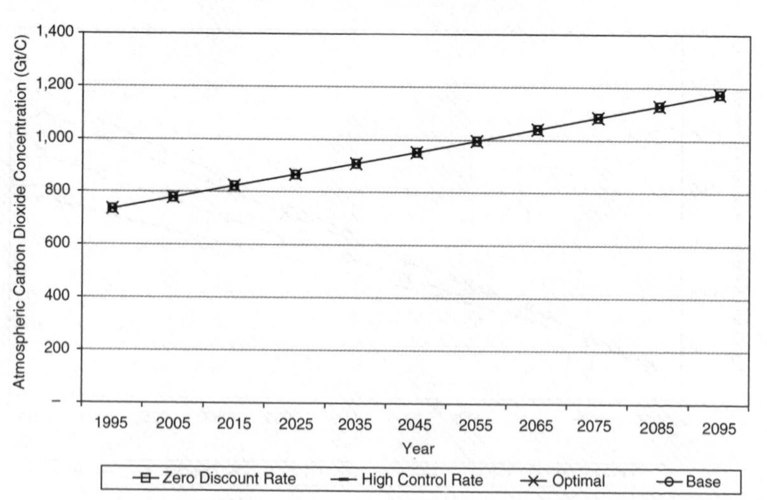

Figure 4.4 South East Asia Industrial Carbon Intensity (metric tons per $US thousand)

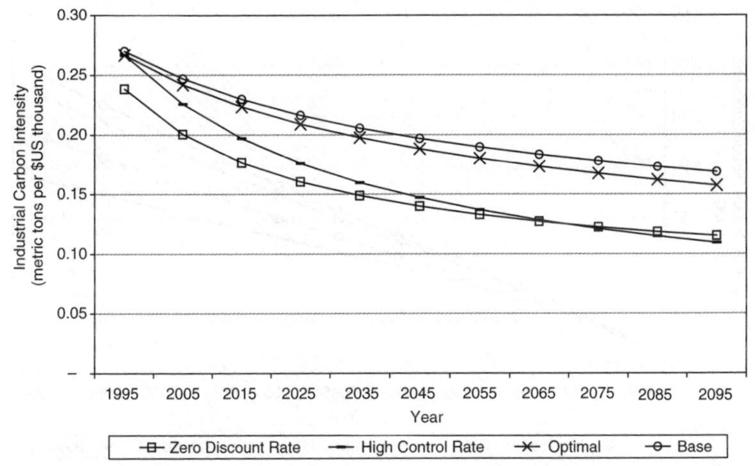

suggests that without international cooperation SEA is unable to influence climate change outcomes by itself through mitigation alone.

Figure 4.4 shows the dynamics of industrial carbon intensity for the five model runs. Not surprisingly the HDRUE model run has a lower carbon intensity because the factors discussed in the paragraph describing Figure 4.2 have forced the carbon intensity of the economy down. The Zero Discount Rate case is lower as well, again suggesting that the altruistic methods forced by that type of discounting have forced carbon intensity down to lower the burden on future generations. Figure 4.5 shows the results of the model runs for industrial emissions, which displays the same pattern as Figure 4.2. Overall, the results of the five model runs have shown reasonable and logical results with respect to the environmental variables based on the structure of the SEADICE model.

4.5.1.2 *Results for the economic variables*

After the discussion of the results of the environmental variables of the SEADICE model it is now time to focus on the economic variables. In Figures 4.6–4.12 it can be seen how the economic variables are affected by the five different model runs. It is apparent that the

Figure 4.5 South East Asia Industrial Carbon Dioxide Emissions (Gt/C per year)

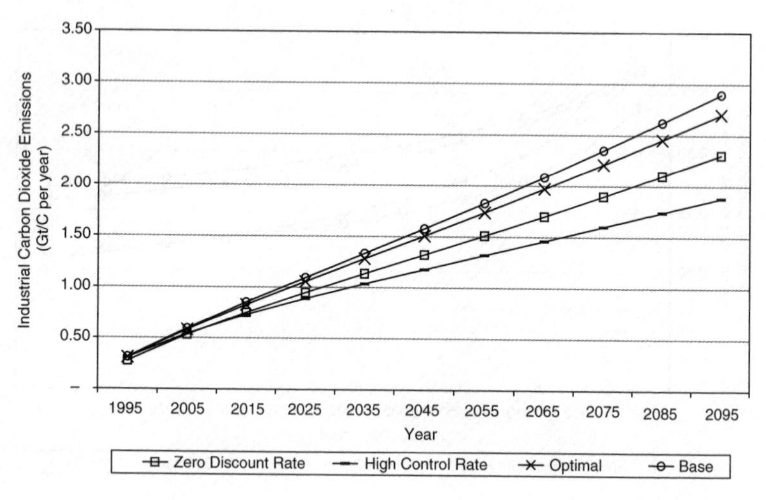

Figure 4.6 South East Asia GDP

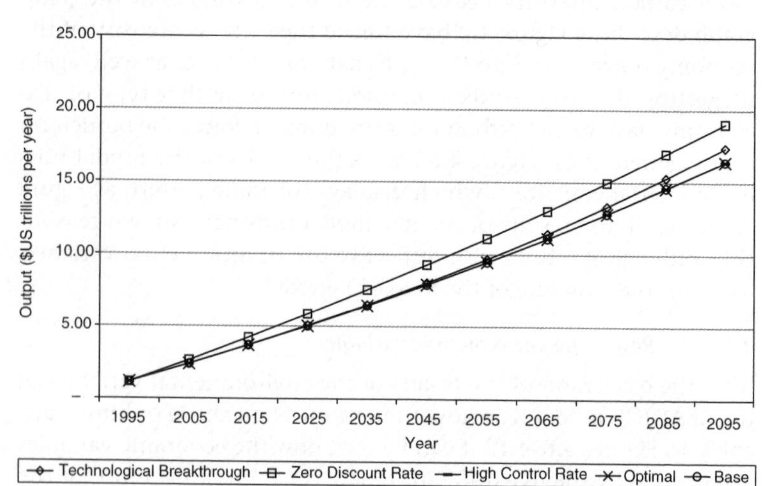

Figure 4.7 South East Asia Capital Stock ($US trillion)

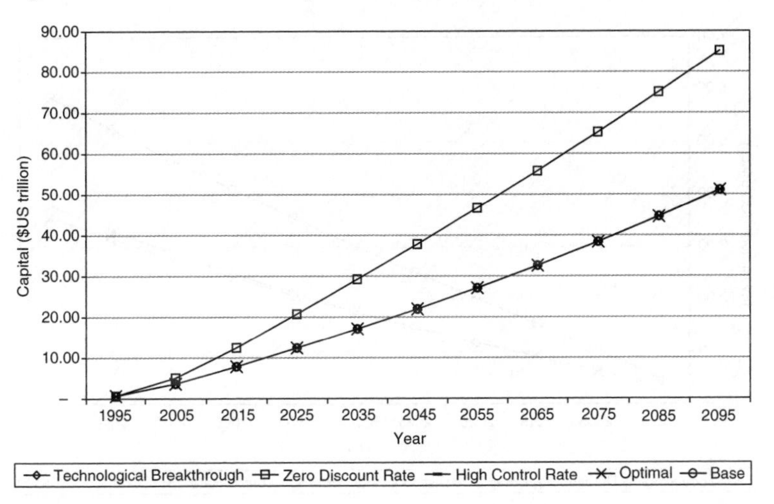

Figure 4.8 South East Asia Consumption ($US trillion per year)

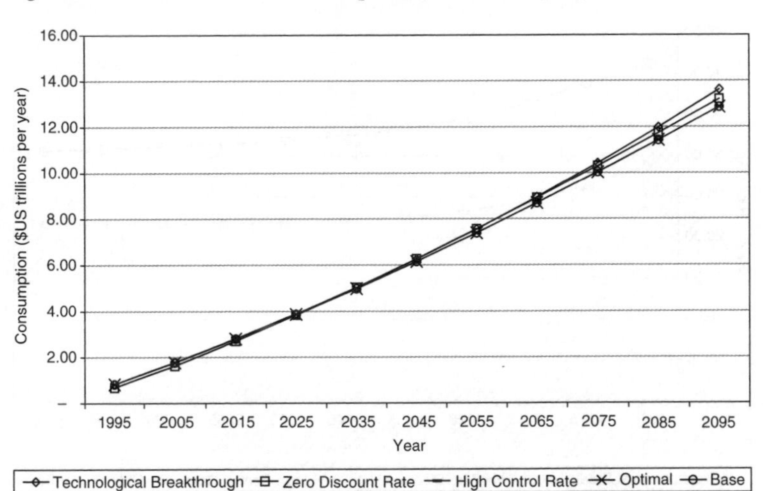

Figure 4.9 South East Asia Investment ($US trillion)

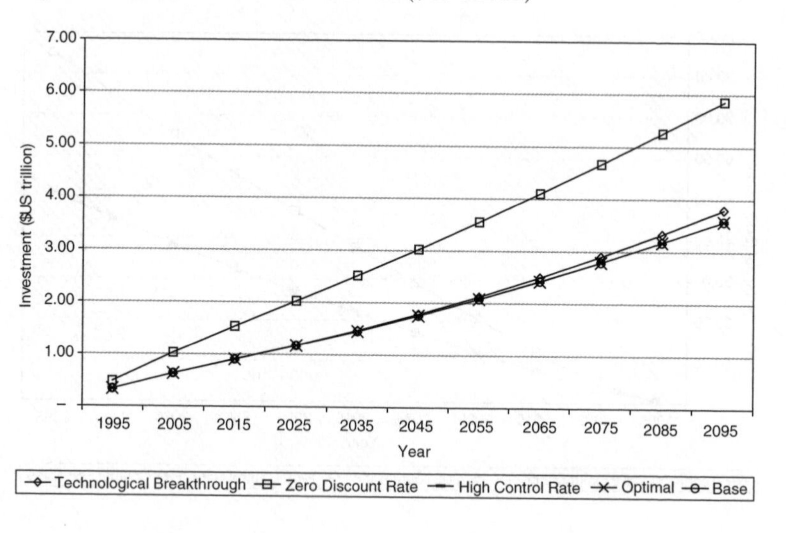

Figure 4.10 South East Asia Saving Rate (%)

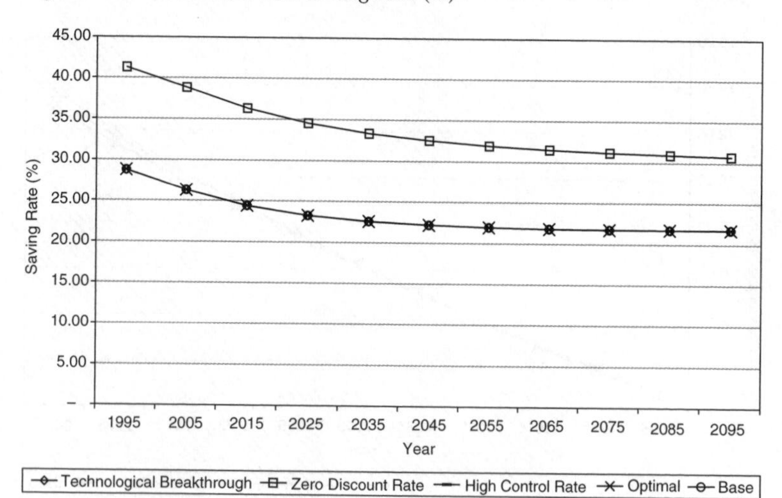

Figure 4.11 South East Asia Consumption Per Capita ($US thousand per year)

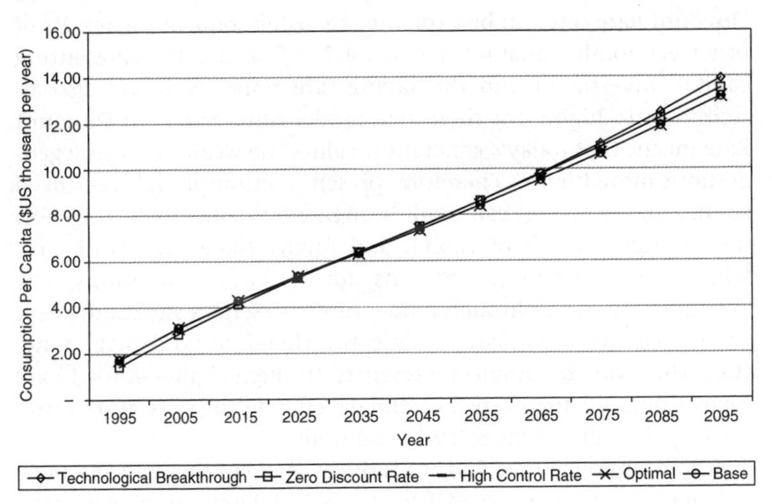

Figure 4.12 South East Asia Interest Rate (%)

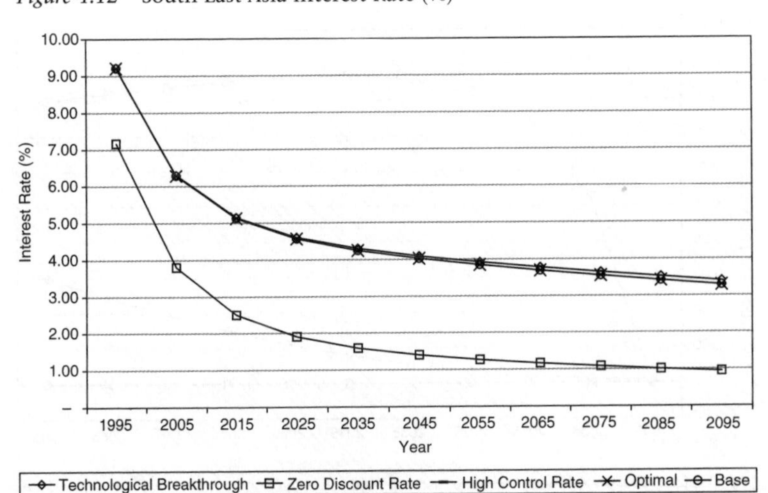

Zero Discount Rate and Technological Breakthrough model runs result in higher levels of GDP up to the 2×CO_2 case. The Zero Discount Rate case can be explained by using some of the results of other economic variables. Figures 4.7, 4.9 and 4.10 representing capital, investment and the saving rate respectively are also all significantly higher for these two model runs. The Zero Discount Rate means that today's generation values the wealth of future generations more highly. Therefore, present consumption is sacrificed so that higher savings can result in higher investment which in turn means higher levels of capital and finally higher levels of GDP which benefit future generations during 2×CO_2 conditions. The Technological Breakthrough model run represents a situation where the model has no climate module and therefore no damage function. Therefore, the results represented through Figures 4.6–4.12 are logical because there are no climate change impacts and consequently they show higher levels of output.

According to Figure 4.13 the sacrifices made in the Zero Discount Rate model run results in a GDP rate 16% higher than the Base case. This indicates that the model is quite sensitive to changes in the discount rate, a conclusion consistent with results from the DICE and ADICE models (Islam 1994; Nordhaus 1994a; Nordhaus and Boyer

Figure 4.13 South East Asia GDP Difference from Base Case (%)

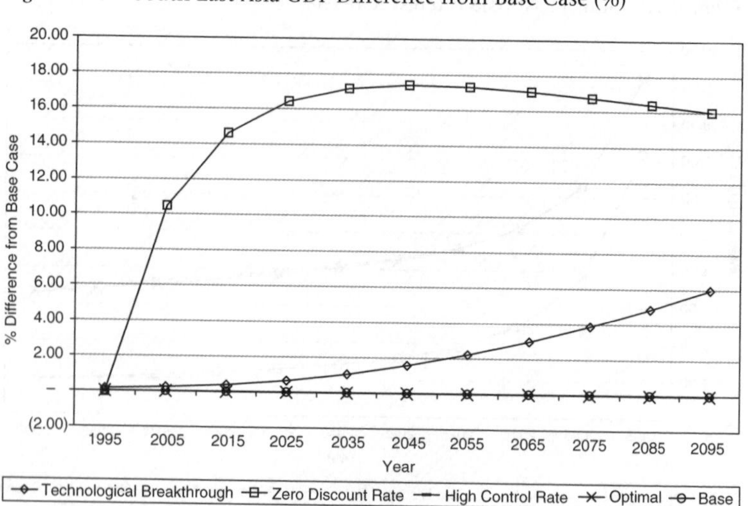

2000). It can also be seen from Figure 4.13 that the Technological Breakthrough model run results in a 6% higher level of GDP at 2×CO_2 conditions. Overall, Figures 4.6–4.13 demonstrate that the SEADICE model is working consistently and therefore has been successfully implemented.

4.6 Optimistic and pessimistic scenarios

The model results representing the different impact scenarios generated in Chapter 3 will be presented in this section. It is apparent from Figures 4.14–4.16 that different scenarios ranging from optimistic to pessimistic make little difference to the results overall. Only three variables are represented here to demonstrate the sensitivity to the scenarios examined, all other variables in the model including environmental and economic, exhibit similar results. This indicates that the model is not very sensitive to changes in the impact parameter on the damage function. The main implication is that the high level of uncertainty behind the estimates made in Chapter 3 become less important. To explain, even order of magnitude discrepancies in the impact estimates for SEA would not have made much of a difference in the overall results at the 2×CO_2 level.[22]

Figure 4.14 South East Asia Investment Scenarios

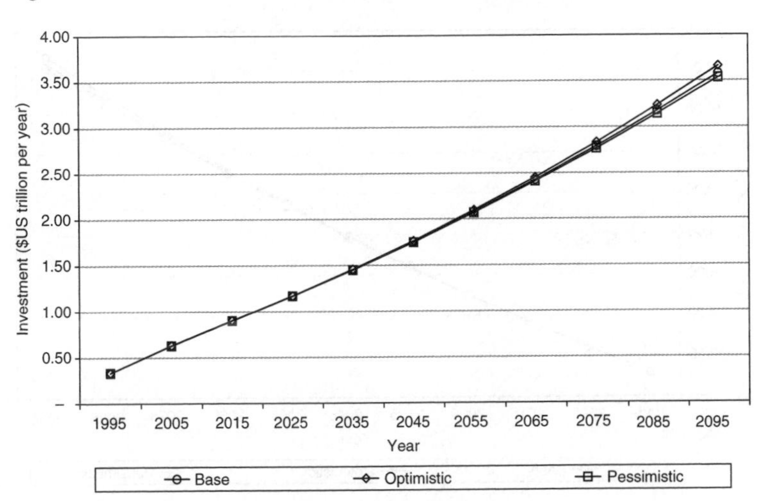

Figure 4.15 South East Asia Consumption Scenarios

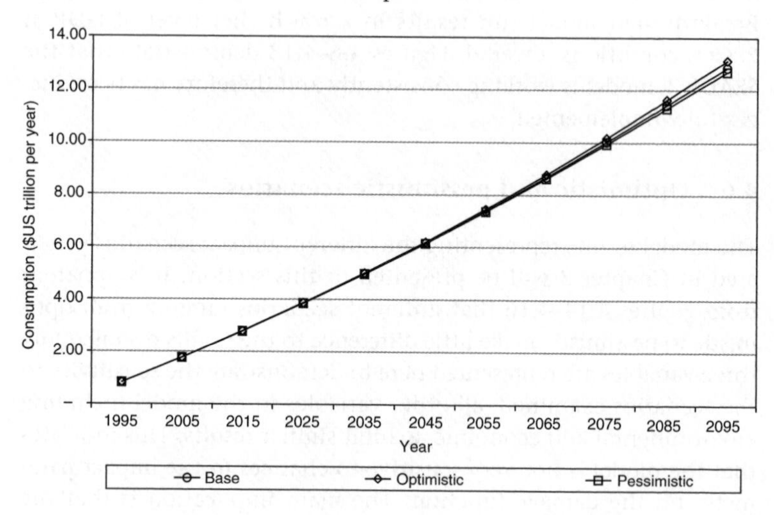

Figure 4.16 South East Asia GDP Scenarios

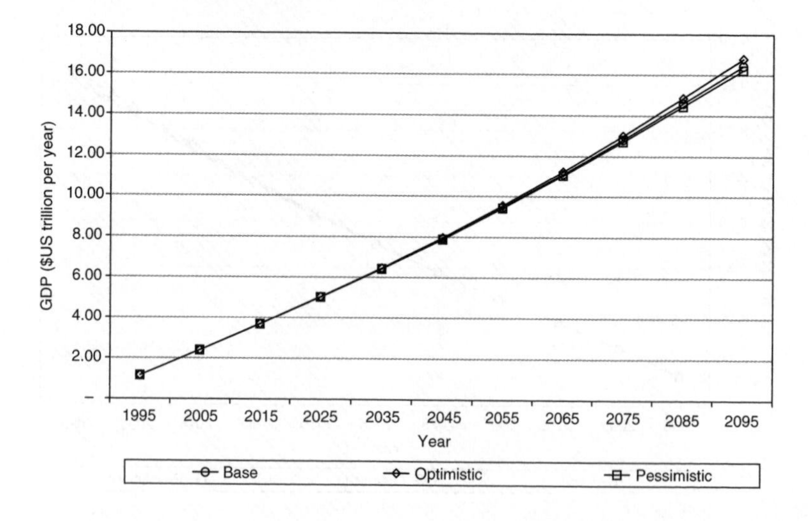

4.7 Conclusion

The implementation of the SEADICE model was the main focus of this chapter. The main finding of the chapter is that the DICE model can be successfully modified to represent the impacts of climate change on a particular region such as SEA. The chapter began with a review of the IAM literature and a discussion of the major modelling issues associated with these types of models. Given several reasons including ease of use, dynamic and optimal results and the already inherent general positive characteristics of IAM models, the DICE model was chosen as the modelling framework upon which to base this book's model. The SEADICE model differs from the DICE model in the respect that many of its parameter values uniquely represent SEA and that the region's emissions are separated from global emissions following the method employed by Islam (1994) for the ADICE model. The solution process for the model was chosen and a detailed explanation of the equations of the model given. The model was solved for five different sets of parameters and also for optimistic and pessimistic impact scenarios. Throughout this process the model performed and produced consistent and logical results. It must be emphasised that the results gained from the SEADICE model are illustrative. Models of this type serve the purpose of expanding knowledge of the relationships between major climate and economic variables, in this case for the SEA region. Now that the SEADICE model has been successfully implemented the possibility exists to further modify the model. In the following chapter this will be done by introducing adaptation into the model.

Appendix 4.A SEADICE (South East Asia DICE) Model

$$\max_{\{c(t)\}} \sum_t U[c(t), L(t)](1+\rho)^{-t}$$

$$U[c(t), L(t)] = \frac{L(t)\{[c(t)]^{1-\alpha} - 1\}}{(1-\alpha)}$$

Subject to: $Q(t) = \Omega(t)A(t)K(t)^{\gamma} L(t)^{1-\gamma}$

$$Q(t) = C(t) + I(t)$$

$$c(t) = \frac{C(t)}{L(t)}$$

$$K(t) = (1 - \delta_K)K(t-1) + I(t-1)$$

$$E(t) = [1 - \mu(t)]\sigma(t)Q(t)$$

$$LU(t) = LU(0)(1 - \delta_1)^t$$

$$ET(t) = E(t) + LU(t)$$

$$M(t) = ET(t-1) + E_{ROW}(t-1) + \phi_{11}M_{AT}(t-1) - \phi_{12}M_{AT}(t-1) + \phi_{21}M_{UP}(t-1)$$

$$M_{UP}(t) = \phi_{22}M_{UP}(t-1) + \phi_{12}M_{AT}(t-1) - \phi_{21}M_{UP}(t-1) + \phi_{32}M_{LO}(t-1) - \phi_{23}M_{UP}(t-1)$$

$$M_{LO}(t) = \phi_{33}M_{LO}(t-1) - \phi_{32}M_{LO}(t-1) + \phi_{23}M_{UP}(t-1)$$

$$F(t) = 4.1\frac{\log\left[\frac{M(t)}{735}\right]}{\log(2)} + O(t)$$

$$T(t) = T(t-1) + \left(\frac{1}{R_2}\right)\left\{F(t) - \lambda T(t-1) - \frac{R_2}{\tau_{12}}[T(t-1) - T^*(t-1)]\right\}$$

$$T^*(t) = T^*(t-1) + \left(\frac{1}{R_2}\right)\left\{\left(\frac{R_2}{\tau_{12}}\right)[T(t-1) - T^*(t-1)]\right\}$$

$$D(t) = Q(t) b_1 T(t)^{b_2}$$

$$TC(t) = Q(t) \theta_1 \mu(t)^{\theta_2}$$

$$\Omega(t) = \frac{1 - \theta_1 \mu(t)^{\theta_2}}{1 + b_1 T(t)^{b_2}}$$

Table 4.A.1 Major Variables of the DICE/SEADICE Model

Exogenous Variables

$A(t)$	=	level of technology
$L(t)$	=	labour inputs (population at time t)
$O(t)$	=	forcings of exogenous greenhouse gases

Parameters

α	=	elasticity of marginal elasticity of consumption
b_1, b_2	=	parameters of damage function
γ	=	elasticity of output with respect to capital
δ_A	=	rate of decline in productivity growth rate
δ_σ	=	rate of decrease in the growth rate of σ
δ_K	=	rate of depreciation of capital stock
δ_{pop}	=	decline rate of population
λ	=	feedback parameter in climate model
ρ	=	pure rate of social time preference
R_1	=	thermal capacity of the upper layer
R_2	=	thermal capacity of deep oceans
$\sigma(t)$	=	GHG emissions/output ratio
τ_{12}	=	transfer rate from upper to lower reservoir
θ_1, θ_2	=	parameters of emissions-reduction cost function
g_{pop}	=	initial population growth rate
g_L	=	initial productivity growth rate
g_σ	=	initial growth rate of σ

Endogenous Variables

$C(t)$	=	total consumption
$c(t)$	=	per capita consumption
$D(t)$	=	damage from greenhouse warming
$E(t)$	=	emissions of greenhouse gases
$LU(t)$	=	emissions from land use change
$ET(t)$	=	total emissions
$F(t)$	=	radiative forcing from GHGs
$\Omega(t)$	=	output scaling factor from emissions controls and climate change damages
$K(t)$	=	capital stock
$M(t)$	=	mass of greenhouse gases in the atmosphere
$Q(t)$	=	gross domestic product
$T(t)$	=	atmospheric temperature relative to base period
$T^*(t)$	=	deep-ocean temperature relative to base period
$TC(t)$	=	total cost of reducing GHG emissions
$u(t)$	=	$u[c(t)]$ = utility of per capita consumption

Policy Variables

$I(t)$	=	gross investment
$\mu(t)$	=	rate of emissions reductions

Table 4.A.2 Initial Parameter Values for the SEADICE Model

α	=	1
γ	=	0.3
b_1	=	0.006288*
b_2	=	2
δ_A	=	4.5
δ_σ	=	2.359
δ_K	=	8
δ_L	=	25.663
$g_L(1995)$	=	20
$g_A(1995)$	=	11
$g_\sigma(1995)$	=	–11.68
K(1995)	=	0.7004*
λ	=	1.41
M(1995)	=	735
$M_{UP}(1995)$	=	781
$M_{LO}(1995)$	=	19230
L(1995)	=	476.74*
ρ	=	3
Q(1995)	=	1.16*
$\sigma(1995)$	=	0.27*
T(1995)	=	0.43
T*(1995)	=	0.06
θ_1	=	0.03
θ_2	=	2.15

* *denotes initial parameter value specific to SEA.*

Appendix 4.B SEADICE Results

Table 4.B.1 SEADICE Climate-Emissions Paths

	Run	1995	2015	2035	2055	2075	2095
Radiative Forcing	1	1.04	1.97	2.84	3.65	4.42	5.16
	2	1.04	1.97	2.84	3.65	4.42	5.16
	3	1.04	1.97	2.84	3.65	4.42	5.16
	4	1.04	1.97	2.84	3.65	4.42	5.16
	5	1.04	1.97	2.84	3.65	4.42	5.16
Atmospheric Temperature (°C above pre-industrial)	1	0.43	0.63	1.03	1.47	1.92	2.35
	2	0.43	0.63	1.03	1.47	1.92	2.35
	3	0.43	0.63	1.03	1.47	1.92	2.35
	4	0.43	0.63	1.03	1.47	1.92	2.35
	5	0.43	0.63	1.03	1.47	1.92	2.35
Atmospheric Carbon Dioxide Concentration (Gt/C)	1	735.00	822.20	909.14	996.40	1,084.78	1,174.39
	2	735.00	822.20	909.14	996.40	1,084.78	1,174.39
	3	735.00	822.20	909.14	996.40	1,084.78	1,174.39
	4	735.00	822.20	909.14	996.40	1,084.78	1,174.39
	5	735.00	822.20	909.14	996.40	1,084.78	1,174.39
Total Carbon Dioxide Emissions (Gt/C per year)	1	0.34	0.87	1.35	1.84	2.36	2.91
	2	0.33	0.85	1.30	1.75	2.22	2.71
	3	0.34	0.87	1.35	1.85	2.38	2.95
	4	0.30	0.77	1.15	1.52	1.91	2.32
	5	0.33	0.75	1.05	1.33	1.61	1.89
Industrial Carbon Dioxide Emissions (Gt/C per year)	1	0.31	0.85	1.33	1.83	2.35	2.90
	2	0.31	0.83	1.28	1.74	2.21	2.70
	3	0.31	0.85	1.34	1.83	2.37	2.94
	4	0.28	0.75	1.13	1.51	1.90	2.31
	5	0.31	0.73	1.04	1.32	1.60	1.88
Industrial Emissions Control Rate	1	–	–	–	–	–	–
	2	1.23	2.78	3.94	4.97	5.94	6.82
	3	–	–	–	–	–	–
	4	11.74	23.16	27.60	29.83	31.13	31.74
	5	1.24	2.51	3.27	3.89	4.44	4.90

Table 4.B.2 SEADICE Economic Variable Paths

	Run	1995	2015	2035	2055	2075	2095
Output ($US trillion per year)	1	1.16	3.69	6.43	9.47	12.82	16.43
	2	1.16	3.69	6.43	9.47	12.82	16.43
	3	1.16	3.71	6.50	9.68	13.32	17.42
	4	1.16	4.23	7.54	11.11	14.97	19.07
	5	1.16	3.69	6.43	9.47	12.82	16.43
Capital ($US trillion)	1	0.70	7.88	17.09	27.08	38.38	51.11
	2	0.70	7.88	17.09	27.08	38.38	51.10
	3	0.70	7.88	17.08	27.07	38.39	51.16
	4	0.70	12.47	29.24	46.65	65.24	85.22
	5	0.70	7.88	17.09	27.08	38.38	51.11
Consumption ($US trillion per year)	1	0.82	2.79	4.98	7.39	10.03	12.86
	2	0.82	2.79	4.98	7.39	10.03	12.85
	3	0.83	2.80	5.03	7.56	10.42	13.62
	4	0.68	2.69	5.02	7.56	10.30	13.20
	5	0.82	2.79	4.98	7.39	10.03	12.85
Investment ($US trillion)	1	0.33	0.90	1.45	2.08	2.79	3.58
	2	0.33	0.90	1.45	2.08	2.79	3.58
	3	0.33	0.90	1.47	2.13	2.90	3.80
	4	0.48	1.54	2.51	3.55	4.67	5.87
	5	0.33	0.90	1.45	2.08	2.79	3.58
Saving Rate (%)	1	28.76	24.40	22.58	21.95	21.76	21.77
	2	28.76	24.40	22.58	21.95	21.76	21.77
	3	28.76	24.39	22.57	21.95	21.77	21.80
	4	41.26	36.29	33.34	31.95	31.21	30.77
	5	28.76	24.40	22.58	21.95	21.76	21.77
Consumption Per Capita	1	1.73	4.28	6.33	8.41	10.67	13.13
	2	1.73	4.28	6.33	8.41	10.67	13.13
	3	1.73	4.30	6.40	8.59	11.08	13.91
	4	1.42	4.13	6.39	8.60	10.95	13.48
	5	1.73	4.28	6.33	8.41	10.67	13.13
Interest Rate (% per year)	1	9.21	5.13	4.25	3.84	3.55	3.29
	2	9.21	5.13	4.25	3.84	3.55	3.29
	3	9.22	5.15	4.30	3.92	3.65	3.40
	4	7.17	2.50	1.59	1.27	1.08	0.94
	5	9.21	5.13	4.25	3.84	3.55	3.29

Table 4.B.3 Different Scenarios – Optimistic to Pessimistic

	Run	1995	2015	2035	2055	2075	2095
Output ($US trillion per year)	Baseline	1.16	3.69	6.43	9.47	12.82	16.43
	Optimistic	1.16	3.70	6.46	9.55	13.01	16.80
	Pessimistic	1.16	3.69	6.42	9.43	12.72	16.23
GDP, difference from reference (%)	Baseline	–	–	–	–	–	–
	Optimistic	0.06	0.15	0.40	0.83	1.45	2.20
	Pessimistic	(0.03)	(0.09)	(0.23)	(0.48)	(0.83)	(1.26)
Consumption ($US trillion per year)	Baseline	0.82	2.79	4.98	7.39	10.03	12.86
	Optimistic	0.82	2.80	5.00	7.45	10.18	13.14
	Pessimistic	0.82	2.79	4.97	7.36	9.95	12.70
Investment ($US trillion)	Baseline	0.33	0.90	1.45	2.08	2.79	3.58
	Optimistic	0.33	0.90	1.46	2.10	2.83	3.66
	Pessimistic	0.33	0.90	1.45	2.07	2.77	3.53
Climate damage as % of output	Baseline	0.15	0.34	0.89	1.82	3.07	4.61
	Optimistic	0.10	0.21	0.55	1.12	1.90	2.85
	Pessimistic	0.19	0.41	1.09	2.21	3.74	5.62

5
Theoretical Discussion of Adaptation to Climate Change and Application to the SEADICE Model

5.1 Introduction

This chapter discusses the concept of adaptation to climate change and then applies it to the SEADICE model to draw some conclusions with regard to the application of adaptation concepts to dynamic optimisation integrated models of climate change economics. It is in essence a form of positive analysis (the purpose of which is to predict or estimate the likelihood) where the key question is 'what adaptations are likely?' This chapter initially provides general descriptions of some of the concepts of adaptation used in the sciences, and ultimately climate change economics before concluding with an application of climate change adaptation to the SEADICE model. In between these topics, the chapter attempts to make three important arguments, that the concept of autonomous adaptation is an important distinction to make, that it is dependent on the level of technology and that economic growth models with endogenous technical progress are suited to representing autonomous adaptation to climate change.

5.2 General adaptation concepts

The word adaptation carries with it variations in meaning for all manner of disciplines.

> The layman, the biologist, the physician, and the sociologist use the word, each in his own way, to denote a multiplicity of

> genetic, physiologic, psychic and social phenomena, completely unrelated in their fundamental mechanisms (Dubos 1965, p. 257).

Even within single disciplines, the concept of adaptation can have several different meanings. Adaptation can take on meanings such as the process underlying natural selection in genetics or the process maintaining social homeostasis in human ecology. The word has at its very core the meaning of adjustment to new circumstances, or to make suitable for a purpose, as the word adapt is Latin for 'to fit'. Hence, in broad descriptive terms, on the one hand adaptation can be reactionary as in adjusting to new conditions, or precautionary with respect to making something suitable for a purpose. While the concept is well developed in the natural sciences, economics is yet to adequately incorporate the concept. The treatment of the concept in the natural sciences will be discussed in the next section.

5.2.1 Scientific concepts of adaptation

The concept of adaptation is most commonly associated with the biological/Darwinian evolutionary doctrines, where since the publication of 'On the Origin of Species' (Darwin 1859) scientists have seriously studied the nature of adaptation as it is related to evolution. Adaptation in a biological sense means the adaptation *over generations* rather than the adaptability of a living organism over its lifetime, although adaptability does determine that individual's chance of reproduction. Adaptability should not be confused with the evolutionary process of adaptation.[23] Adaptation subsumes four processes: adaptation by natural selection; physiological or behavioral plasticity during the developmental process; behavioral choices that enhance individual welfare; and corporate behavioral choices that benefit the group. In other words adaptation has been considered from four perspectives over time: genetic, physiological, behavioral and cultural (Ulijaszek 1997).

Adaptation can take on different meanings between and even within scientific disciplines. Some examples include Baker who defines an adaptation as '...simply any biological or cultural trait which aids the biological functioning of a population in a given environment' (Baker 1984). Whereas, Frisancho defines adaptation as 'any change in an organism resulting from exposure to an altered environment that enables the organism to function more efficiently in the new environment...' (Frisancho 1993, p. 486), but 'it is applied to all levels of biological organization from individuals to populations' (Frisancho 1993, p. 4). Marks (1995) defines adaptation as the process by which a feature providing a benefit over its alternatives to an individual in a particular environmental circumstance arises. This variance in views on adaptation makes it clear that even though the concept is relatively poorly represented in economics (i.e. compared to biology) it is still substantially conjectural in other disciplines.

Recent work on adaptation concepts done in the sciences but which has applications to human social systems (including economic systems) include some interesting work being done by a team at the Santa Fe Institute on the concept of Complex Adaptive Systems (CAS). A CAS is a system that is made up of a large number of active elements, such as a forest, large city, ant nest or central nervous system (Holland 1995). The theory states that each CAS should have the same underlying rules (several mechanisms and properties) that govern its adaptive behavior. The only major difference between each CAS is the time frame in which adaptation takes place, a forest may take years or decades to adapt to changes in the environment whereas a central nervous system may operate within the time frame of seconds or minutes. This type of research has the potential to be important for many areas of science as it may reveal universal relationships with respect to adaptation that can be applied across disciplines including economics.

5.2.2 Economic concepts of adaptation

The human is a slow-breeding species capable of only slow genetic transformation, genetics cannot explain the rapid advance of our species, instead the primary force of adaptation has been our ability to create technology, goods, services and stores of knowledge. Adaptation in economics is not a function of passing genes through successive generations but rather the process of learning and storing knowledge so that others can use it now and in the future. Humans have usurped many of the biological mechanisms for adaptation with the ability to change the physical environment in which they live. This is different in a spatial sense from many scientific notions of adaptation in that it is possible to pass on the information needed for the adaptation across populations within a generation and also across generations. In the natural sciences adaptation concepts are exclusively couched in terms of reactions to change, whereas in the social sciences it is possible for adaptation to be precautionary.

In mainstream economics the term adaptation is not commonly used to explain any type of economic relationship. If adaptation is considered, it is as a basic and very broad concept in economics, in which it is referring to the most basic behaviors in economic theory. The seller in a market who adjusts his price in reaction to an increase in demand is in essence adapting to changed circumstances. However this example of adaptation is too general to be of any use for an application to climate change. Ideally, climate change adaptation needs to be conceptualised in terms of an economic unit adjusting to exogenous events in an imperfect world, with uncertain outcomes. These types of adaptation concepts have been used in some areas of economics. Most notably the concept has been embraced by evolutionary economics. This is explored in more detail in Section 5.4. In the meantime an attempt is made in the following section to provide a theoretical basis for adaptation to climate change.

5.3 Adaptation concepts for climate change

Possibly the first reference connecting climate change and adaptation can be attributed to Darwin (1859) where in 'On the Origin of Species' the point is made that climate is the most important 'check' that exists in determining the population of a species. It is explained that the action of the climate increases the severity of the struggle for existence and therefore accelerates evolution (adaptation). The conceptual foundations of adaptation to climate change have been developing over the last decade (Feenstra et al. 1998; Dixon 1999; Fankhauser, Smith and Tol 1999; Mendelsohn 2000).

In Feenstra et al. (1998) it is stated that the clearest conceptual treatments of climate change adaptation are in Chapter 7 of the SAR of WGIII, and in the IPCC Technical Guidelines for Assessing Climate Change Impacts and Adaptations (Carter et al. 1994). However, this conceptual framework was surpassed by the documentation available from the 1998 IPCC conference on climate change adaptation (Dixon 1999). Since then the TAR of the IPCC has become the new international benchmark (IPCC 2001a, 2001b, 2001c). In this section a further attempt is made to advance the notion of economic adaptation to climate change.

Tol, Fankhauser and Smith (1998) examine the state of the art in sectoral adaptation studies and find four types of approaches to adaptation that are covered in the literature. Aside from no adaptation, where it is assumed that humans are passive in the face of climate change, which is useful as a reference point, these are: Arbitrary adaptation, where adaptation levels are selected arbitrarily by economic agents; Observed adaptation, where historical spatial and temporal analogues are used to examine how societies have adapted to past climate variability; Modelled adaptation, which involves the use of behavioral models to predict the adaptive behavior of economic agents and is used mainly to model autonomous adaptation; and Optimal adaptation, which is based on economic efficiency and economic agents who equate their marginal benefits and marginal costs of climate change. Tol, Fankhauser and Smith (1998) makes the point that no comprehensive study of optimal adaptation has been undertaken. Later in this chapter an attempt is made to represent optimal adaptation (at least illustratively).

5.3.1 Definitions of climate change adaptation

The closest there is to an 'official' definition of adaptation to climate change is in Chapter 18 of the WGII document of the TAR of the IPCC (2001b) where adaptation is defined as:

> Adaptation refers to adjustments in ecological, social, or economic systems in response to actual or expected climatic stimuli and their effects or impacts. It refers to changes in processes, practices, and structures to moderate potential damages or to benefit from opportunities associated with climate change. (p. 642)

This definition is very similar to that given in the equivalent IPCC document in 1996,[24] where in simpler terms the definition states that adaptation is any response to predicted or actual climate change. An alternative definition appears in Smith et al. (1996, p. vii), where adaptation to climate change is defined as 'all adjustments in behavior or economic structure that reduce the vulnerability of society to changes in the climate system.' This definition is more restricted in the sense that only reductions in vulnerability are considered adaptation. Therefore, any policies that attempt to reduce the effects (mitigation) of climate change are not considered. Another definition is found in Fankhauser (1998, p. 3) where adaptation is defined as 'projects and policy measures that are undertaken to ease the adverse impacts of climate change'. Feenstra et al. (1998) states that Fankhauser's definition is an institutional definition of adaptation, whereas an economic definition would be broader, including actions taken by economic agents and governments to learn about climate change and distribute this information, and to reallocate resources in an efficient manner to adjust to the negative impacts of climate change. One of the main reasons that adaptation has not been effectively dealt with in the climate economic impacts literature is that economics itself has found it difficult to deal with the concept of adaptation. However, broad definitions of adaptation such as those just shown should not be the ultimate goal and are less relevant as a guide for economic research because adaptation to climate change can be split into distinct types that provide clearer conceptual guideposts for researchers.[25]

5.4 Autonomous and planned adaptation

This section attempts to provide a more detailed conceptual framework upon which further studies of climate change adaptation can be based. One of the most important factors in the concept of adaptation to climate change is the acknowledgement of autonomous adaptation and planned adaptation as separate types of adaptation (Carter et al. 1994; UNEP Collaborating Centre on Energy and the Environment 1998; Dixon 1999; Leary 1999; IPCC 2001b). This distinction along with explanations of each climate change adaptation type will be done later in this section. A weakness with many of the studies that have examined adaptation to climate change is that the distinction between autonomous and planned adaptation is not taken into account or is poorly structured. It is stated by IPCC (1996b) that almost all case studies do not address autonomous adaptation in any way; this is still the case today. Ignoring this important distinction can make any general discussion and quantification of adaptation to climate change largely irrelevant as significant distortions may exist in any results obtained from a definition of adaptation that is too simple. The distinction between autonomous and planned adaptation is important because both have significantly different characteristics that have implications for climate change impacts and mitigation. Factors such as the distinction between private (autonomous) and government (planned) actions are also useful in allowing adaptation to be more accurately modelled for economic systems.

5.4.1 Autonomous adaptation

Autonomous adaptation can be broadly defined as those responses to changes in information or physical effects of climate change that occur automatically as a result of economic agents maximising welfare. Authors such as Leary (1999) define autonomous adaptation as the responses to climate change that economic agents choose when acting autonomously.[26] The extent of autonomous adaptation is determined primarily by an economic agent's ability to respond to change which is limited by factors such as time, available resources and willingness to respond. Leaving markets alone to adapt to climate change is not generally regarded as a viable policy option as many countries would not find it possible to fully adapt to climate changes due to inhibiting factors such as high natural vulnerability to climate and the inability of market mechanisms to identify and react to changes in climate (IPCC 1996b). This is important for the economic analysis of climate change impacts because it acts as a base for any estimation of climate change damage (IPCC 2001b). This issue will be explored in the following section.

5.4.1.1 The importance of autonomous adaptation for base case estimates

The estimation of autonomous adaptation is extremely important for providing base levels upon which overall adaptation benefits and consequently climate change impacts can begin to be measured. Statements made in IPCC (2001b) support this, such as '... to assess the dangerousness of climate change, impact and vulnerability assessments must address the likelihood of autonomous adaptation' (p. 881). This is illustrated more clearly by the following table adapted from the UNEP Collaborating Centre on Energy and the Environment (1998).

Table 5.1 Scenarios for Estimating Adaptation Costs and Benefits

	Existing Climate	Altered Climate
No Adaptation	***Business as Usual*** *Counterfactual* – what would occur if climate did not change	***Dumb Farmer*** *Counterfactual* – what would occur if climate change happened but there was no adaptation
Autonomous Adaptation	***Dumb Engineer*** *Counterfactual* – what would happen if climate did not change, but people decided to adapt	***Optimal Outcome*** *Predicted* – what would happen when climate changes and autonomous adaptation occurs

Source: Adapted from UNEP Collaborating Centre on Energy and the Environment (1998).

Only two cells in Table 5.1 represent optimal outcomes, the top left box (Business as Usual) and the bottom right box (Optimal Outcome). The top right box and bottom left box represent the Dumb Farmer (no action with reason to act) and Dumb Engineer (action with no reason to act) scenarios respectively. To measure the impact of climate change one must compare the costs and benefits of the Dumb Farmer scenario with the Optimal Outcome scenario. This is necessary to isolate the benefits and costs associated solely with adaptation from the combined effects of climate change and adaptation. However, at the moment most studies are estimating climate change impacts as the difference between the Business as Usual scenario and the Dumb Farmer scenario. That method has the potential to bias negative impact estimates upwards. At this stage it is not known how significant this bias might be. The following diagram further illustrates this problem (Figure 5.1 adapted from Feenstra et al. 1998).

Imagine the lowest line sloping upwards from left to right in Figure 5.1 is the marginal cost curve for a climate sensitive product such as a type of agricultural output. The marginal cost curve shifts to the left as a result of climate change increasing costs at every level of output. The leftmost curve represents the scenario where climate change occurs and no adaptation takes place (Dumb Farmer sce-

Figure 5.1 Costs of Climate Change Adaptation

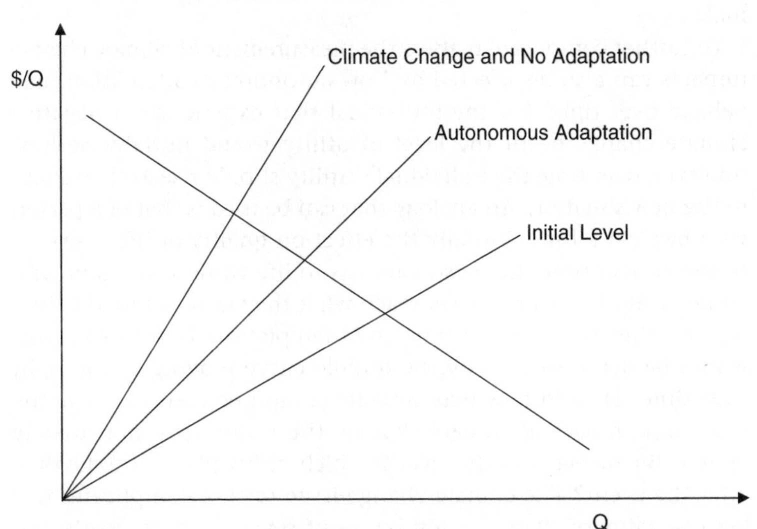

Note: Q = output

nario). The middle curve represents the scenario where climate change occurs and autonomous adaptation takes place. The importance of including autonomous adaptation can be seen from Figure 5.1. If autonomous adaptation is not taken into account the impacts of climate change can be overstated, (the difference between the middle curve and the leftmost curve) depending on the likely level of autonomous adaptation. 'Neglect of the issue of induced technical change and other adaptive responses may invalidate the policy implications drawn from most integrated assessment models developed to date' (Grubb, Chapuis and Duong 1995, p. 417). However, the level of autonomous adaptation has proven very difficult to estimate and has generally been ignored in economic studies.

The opportunity for planned adaptation is determined by the level of damage remaining between the initial scenario and the autonomous adaptation scenario. If planned adaptation policies are to be implemented based on economic cost then an accurate estimate of the potential benefits and costs of each policy are needed, which includes an accurate assessment of the impacts of autonomous adaptation. As can be seen from the preceding argument, autonomous adaptation is an important distinction to make.

To further complicate matters the measurement of climate change impacts can also be affected by how autonomous adaptation may behave over time. For the individual that experiences a negative climate change event the level of utility would initially be low, however, over time the individual's utility should rise as they adapt to the new situation. An analogy that can be used is that of a person who has lost a limb, initially the effect on quality of life is severe, however over time the person adjusts to life without the limb and while utility levels may never reach what they were before the limb was lost they will rise over time and then plateau. In Figure 5.1 this would be demonstrated by the middle curve moving to the right over time. Therefore, which measure should be used to gauge the economic impact of climate change; the utility loss immediately after a climate event or the level at which utility plateaus some time after the event? The climate change issue further complicates this because climate change is a mixture of severe impact events and gradual changes over time. If a measure of utility is taken immediately after an event then mitigation policies would be more cost effective relative to adaptation policies because the relative benefit of prevention would be higher. However, if the level of utility is measured with the inclusion of gradual adaptation over time then adaptation policies become more cost effective. This particular issue is beyond the scope of this book, but is an interesting avenue for further research in the climate change impacts literature.

5.4.1.2 Technology as a determinant of autonomous adaptation

In theory the level of autonomous adaptation should be largely dependant on the adaptive capacity of the economy. According to IPCC (2001b), the adaptive capacity of the economy to climate change is determined by economic wealth, technology, information and skills, infrastructure, institutions and equity. These factors are by no means independent of each other, nor are they mutually exclusive. In this book the level of technology is considered an important determinant of autonomous adaptation. Indeed, IPCC (2001b) states that 'Many of the adaptive strategies identified as possible in the management of climate change directly or indirectly involve technology' (p. 896). Support for this type of argument already exists in the climate change impacts literature, as it is generally regarded that more advanced economies will be less vulnerable to the effects of climate change due to several factors such as a lower exposure of climate sensitive industries and more advanced technology (Sheraga and Grambsch 1998). Technologies not available to everyone, such as irrigation, storm early warning systems and drought resistant crops enable those economies, which have access to advanced technologies to cope with climate change more easily. Therefore, it is assumed in this book that technology can be considered a determinant of the level of autonomous adaptation to climate change.

5.4.2 Planned adaptation

Planned adaptation is that which occurs in anticipation of the effects of climate change in order to reduce vulnerability. Planned adaptation will be needed for two main purposes, the first is to protect public assets where benefits are external to private economic agents and consequently adaptation requirements need to be identified and implemented. The other major need for planned adaptation occurs in situations where it is determined that private economic agents require assistance in protecting private assets, in other words, to facilitate autonomous adaptation (Leary 1999). Planned adaptation may involve dissemination of information regarding climate change and variability, or the provision of laws, subsidies or taxes designed to provide incentives for the protection of private economic assets.

In order to illustrate the difficulties involved in the implementation of planned adaptation an examination of how the timing of planned adaptation decisions might work in a cost-benefit framework is undertaken.[27] According to cost benefit theory, investment should be made over time up to the point where the benefits of delay are greater than the costs. To illustrate, assume a planned adaptation project named a with costs C^{α} in period 0, results in unaffected damages of d_0 in period 0 and a stream of future reduced damages (benefits) of $d_1 \ldots d_n$.

Therefore, given a discount rate of r, the net present value (NPV) of this planned adaptation investment would be:

$$NPV = C^{\alpha} + d_0^{\alpha} \frac{d_0^{\alpha}}{(1+r)} + \frac{d_2^{\alpha}}{(1+r)^2} + \ldots + \frac{d_n^{\alpha}}{(1+r)^n}$$

If the adaptation project were postponed for one period there would be unaffected damages for periods 0 and 1 and reduced damages for periods 2 onwards. Assuming the new costs are C^b there would be a scenario where delay is preferable if:

$$C^{\alpha} - \frac{C^{\beta}}{(1+r)} > (d_0^{\beta} - d_0^{\alpha}) + \frac{(d_1^{\beta} - d_1^{\alpha})}{(1+r)} + \frac{(d_2^{\beta} - d_2^{\alpha})}{(1+r)^2} + \ldots + \frac{(d_n^{\beta} - d_n^{\alpha})}{(1+r)^n}$$

In other words if the cost savings of the delay (the left hand side of the equation) is larger than the increase in damage from the delay (the right hand side of the equation) it will be logical to delay planned adaptation.

In a real world application of this theory to the climate change adaptation problem the cost side of the equation would be much more certain than the benefit side. The costs are dependant upon factors largely under the control of the policy maker or investor. However, the potential benefits are largely out of the control of the investor. While the investor may know with certainty the initial level of damage; from then on it is uncertain as it relies upon two things: the extent of climate change (which is uncertain) and the impact of climate change which relies on knowledge of the amount of autonomous adaptation (which is yet to be adequately estimated in the literature). The longer the timeframe for the investment the more problematic this underlying uncertainty becomes. The information underlying the forecasts of climate change, and the subsequent level of damage are vital for the facilitation of anticipatory investments in climate change adaptation.[28] Therefore, until reliable impact estimates can be made it can be assumed that the implementation of planned adaptation will be mostly reliant upon governments that are willing to invoke a form of precautionary principle with respect to adaptation to climate change. However, planned adaptation cannot rely on the precautionary principle in the same way that mitigation policies do. Mitigation policy is based on the precautionary principle, where policies are chosen on a least cost basis given the desired level of GHG reduction. Their effectiveness can be measured now by the amount of emission reduction achieved. However, a reliable measure by which least cost adaptation policies can be measured is yet to be developed. It will take a great leap of faith at this time for policy makers to implement adaptation policies based on some form of precautionary principle.

Government intervention will be needed for significant levels of planned adaptation to occur. Planned adaptation will be a required response from governments to counter the damaging effects of climate change that will not be accounted for through autonomous means. Planned adaptation is precautionary in nature and relies predominantly on the uncertainty of future forecasts of climate change and the value society places on future generations. As a result of the uncertainties involved it is quite possible that planned adaptation policies may not occur for quite some time. In a more theoretical sense, planned adaptation could be defined as all adaptation strategies that are justified by cost benefit analysis that are not covered by autonomous adaptation adjustments. Therefore, as discussed in the previous section, in order to design planned adaptation policy options the amount and types of autonomous adaptation that will take place needs to be known.

5.4.3 Summary definitions

In economic modelling it is important to simplify relationships so that they can be better understood. As a result of this, based on the discussion thus far in this chapter a simple definition for autonomous and planned adaptation that would enable them to be more easily understood is given. Autonomous adaptation is defined as the adjustment *an economic agent* makes that occur primarily in *reaction* to changes in that agent's environment.[29] Planned adaptation is defined as being the adjustments *an economic group* makes that occur in *anticipation* of changes in that economic group's environment that will not be taken into account by autonomous adaptation. In very simple terms autonomous (endogenous) adaptation occurs primarily in a reactive way at an individual level, whereas planned (exogenous) adaptation occurs primarily in an anticipatory way at a group level. The information required for autonomous adaptation to occur is minimal whereas it is substantial for planned adaptation. Table 5.2 provides a summary of the general characteristics of autonomous and planned adaptation to climate change.

Table 5.2 General Characteristics of Climate Change Adaptation

Planned Adaptation	Autonomous Adaptation
• Exogenous	• Endogenous
• Government	• Private
• Country Level/International	• Local
• Group	• Individual
• High uncertainty	• Low uncertainty

5.5 Endogenous technical change: a representation of autonomous adaptation

In the previous section it was established that the technological capabilities of an economy are a determinant of the amount of autonomous adaptation. Therefore, representing technology accurately in an economic growth model should improve that model's representation of autonomous adaptation. Ausubel (1995) states that the climate change literature mostly underestimates the importance of technical change for mitigation and adaptation to climate change. The latest developments in the representation of technology in economic growth theory can be described by two separate disciplines, evolutionary economics and new growth theory. Both of these disciplines share the characteristic that they attempt to portray technological progress as a process that occurs endogenously within an economy (Mulder, Reschke and Kemp 1999; Maurseth 2001). This section will discuss the role of technological progress in theories of economic growth and review the two disciplines mentioned in order to provide a theoretical basis for the application of endogenous technical progress to the SEADICE model attempted later in this chapter.

5.5.1 Theories of economic growth and the emergence of endogenous technical progress

While ideas about the character of economic growth have been around for over 200 years, formal mathematical modelling of economic growth is relatively recent (Sheehan 2000). The seminal optimal growth model was Frank Ramsey's 1928 work, 'A Mathematical Theory of Saving'; other works of importance at the beginning of formal growth models were John von Neumann's (1938) 'A Model of General Equilibrium', and Roy Harrod's (1939) 'An Essay in Dynamic Theory'. From these beginnings the neoclassical theory of economic growth emerged with the contributions of Swan (1956) and Solow (1957). The neoclassical theory of economic growth demonstrated that in a purely competitive economy steady state economic growth was possible. Neoclassical models are characterised by well behaved production and utility functions and rational and optimising behavior in a freely operating market with competitive equilibriums in capital and labour markets. In particular the production function exhibits constant or decreasing returns to scale, as well as achievement of the Inada conditions.[30] The motivation here is not to provide the exact definition of neoclassical growth models, of importance to this book is that these models featured *exogenous* technical change as the main driver of long term, steady state per capita growth.

The conclusion that firms fully anticipate the consequences of introducing new technology (through exogenous technical progress) and that, therefore the learning processes of firms can be ignored, became over time inconsistent with the characteristics of economic growth. Over time it became apparent that it was not possible to explain a significant proportion of the increase in output per worker in developed countries using the traditional neoclassical growth model. Sen (1970) provides samples where seven-eighths of the observed increase in output per worker in the United States through 1909–49 could not be explained by increases in the capital-labour ratio alone. Therefore, the vast bulk of economic growth resided in the exogenous residual of the neoclassical growth model and could not be explained.

This particular problem was one of the reasons for the subsequent attempts at endogenising technical progress in economic growth models. Barro and Sala-i-Martin (1995) provide a comprehensive survey of the endogenous technical change literature. Specific examples of some of the more important early theoretical attempts can be found in Haavelmo (1954); Kaldor and Mirrles (1962) and Arrow (1962). However, the particular emphasis of this book rests on the developments in the field of new growth economics (Cortright 2001). New growth theory has the particular trait of describing technological progress as the result of conscious profit-motivated investment decisions. An early attempt at analysing endogenous profit-motivated technological investments was made by Shell (1966) who demonstrated how public investments in technological progress might contribute towards economic growth. Later still, came the origins of the new growth theory, with the models of Romer (1986, 1990) and Lucas (1988). The fundamental basis of new growth economics is explaining the role that technology (knowledge) plays in economic growth. New growth theory allows sustained economic growth, but with increasing returns to scale in the aggregate production function. Although this literature is diverse, overall it allows for more improved explanations for phenomenon such as the continued divergence in economic growth and development between some economies that neoclassical models could not explain (England 1994; Sheehan 2000).

The other main branch of economic growth theory is evolutionary economics which has grown rapidly as described by Hodgson (1997), largely as a consequence of the seminal work of Nelson and Winter (1982). Evolutionary economics provides insights into how markets function; how innovations initiate and technologies change. The word evolutionary is currently being applied to a wide array of economic approaches (Magnussan 1994). It would be a mistake in this book to apply the term evolutionary economics in a general way assuming that an overall accepted meaning exists. Work using the term 'evolutionary' includes theory influenced by authors such as Veblen, Schumpeter, Marshall, Marx, Smith and new strands built upon complex mathematics such as neural networks, replicator dynamics and genetic algorithms. This encompasses a wide range of views. Hence, there is no consensus on what the term evolutionary economics should mean (Hodgson 1997). Even evolutionary economists agree that no single, comprehensive definition exists for the characteristics of economic evolutionary change (Radzicki and Sterman 1994). While many areas of evolutionary economics offer substantial promise for particular aspects of climate change economics, and in particular adaptation, this book will use theories arising from new growth economics, as its methods are more amenable for application to a model based on the neoclassical theory of growth such as SEADICE.

Models incorporating endogenous technical progress have been created in subject areas related to this book. Recently, models featuring endogenous technical progress have been applied to environmental policy (Messner 1997; Smulders 1998). A comprehensive survey of technical progress studies in climate change can be found in Azar and Dowlatabadi (1999). The possible role of endogenous technical progress in influencing GHG rate of control have been covered in Peck and Teisberg (1995), Grubb, Chapuis and Duong (1995), Hall and Mabey (1995), Nordhaus (1997), Edmonds, Roop and Scott (2000) and Janssen and De Vries (2000). However, the issue in these studies has been couched in terms of the effect of mitigation policies on the rate of technical change; the connection between adaptation and technical change has not been made.

The neoclassical theory of economic growth failed to represent technical change adequately, consequently attempts at endogenising technical change emerged as an attempt to improve upon the neoclassical model. The two main schools of thought emerged, new growth theory and evolutionary economics. Although both of these two research streams do offer possibilities for the representation of autonomous adaptation to climate change, new growth theory is chosen for the purposes of this book as it is more amenable to the SEADICE type of model. Before applying endogenous technical progress to the SEADICE model it will be useful to first review some of the attempts that have already been made to represent adaptation to climate change in economic models.

5.6 A review of the modelling of climate change adaptation

In this section the attempts that have already been made on the economic modelling of adaptation to climate change will be reviewed. Over time economists have come to realise that adaptation must be taken into account with any comprehensive attempt at modelling the effects of climate change (Fankhauser 1996). The incorporation of adaptive or evolutionary concepts in economic models has already been demonstrated to be difficult in this chapter. The following review will highlight the attempts that have been made thus far to incorporate adaptation.

5.6.1 Review of climate change models incorporating adaptation

Since SEADICE is an IAM this book will review the application of adaptation to this particular type of model. There are many different types of modelling covering the impacts of climate change apart from IAMs, however, a more focused literature review will prove to be more relevant for this book. The need for economic analysis of the climate change problem has created a range of IAMs that have been developed in the past decade. As explained earlier in the book an IAM is generally regarded as any model that integrates multiple disciplines within its framework. In the case of climate change the models are generally hybrids of economics and the natural sciences which predict climate change and the consequent economic impacts. The main focus here will be on those IAMs that explicitly account for climate change adaptation.

5.6.1.1 Modelling of adaptation in climate change economics

The current approach of climate change modelling is to either exclude adaptation, or to arbitrarily choose a level or representation of adaptation and to monitor its effect on the results. The latter has been done predominantly in static general equilibrium models. Dynamic models such as DICE and ADICE have not explicitly accounted for adaptation. In fact of the many IAMs that have been developed over the last decade only the PAGE model (Hope, Anderson and Wenman 1993; Plambeck, Hope and Anderson 1997), and the model developed by Janssen and De Vries (2000) represent adaptation explicitly as a policy variable.

The PAGE model (Policy Analysis of the Greenhouse Effect) is a simulation model with wide ranging climate change impact assessments. The PAGE model represents adaptation as a situation where society fully adapts to climate change at no cost for low levels of climate change. However, when a threshold is reached damage costs begin to be incurred. Therefore, the way this model interprets the effect of adaptation is by seeing what happens to the model output when the threshold level is changed. This threshold level is represented by three single values for each scenario of adaptation. The first value increases the slope of the tolerable profile (that gives the maximum rate of change in temperature tolerable before some damage occurs) in an impact sector, the second factor increases the plateau parameter (which gives the maximum absolute temperature change tolerable) in an impact sector, the third value describes the percentage decrease in impact in an impact sector if the change in temperature exceeds the tolerable limits. Therefore, in order to represent the effects of an adaptation policy the damage function can be changed to represent a different level of vulnerability to climate change. When this is done the cost of adaptation is proportional to the changes in these three variables. The results of PAGE demonstrate that the costs of adaptation are estimated to be around one-third of the benefits from the reduction in economic impact, and therefore is justified as a viable policy option. This attempt at representing adaptation rests on an assumption that a level of climate change exists where full adaptation suddenly stops. The theory is that for an economy the effect of changes in vulnerability (the ability to adapt autonomously) can be represented by changes in the threshold profile. Unfortunately, there is no explanation of why the profile might change. Technical progress is exogenous in this model and plays no part in the impacts of adaptation.

The model developed by Janssen and De Vries (2000) is another simulation model with an economic component based on the neoclassical model of Nordhaus (1994a). Janssen and De Vries (2000) models the adaptive behavior of economic agents when presented with expectations and actual outcomes of climate change using the method called Genetic Algorithm (Holland 1995) which is inspired by biological concepts. In the model there are three types of world view that shape the actions of economic agents. It is highlighted by the authors that the model is for illustration purposes only and only implementation in more detailed simulation models will be adequate for climate change policy relevance. The model is populated by a substantial number of agents who adhere to a mix of the three principles but can change their preference according to the matching of their observations and expectations. The real world is observed by the agents by a few indicators (atmospheric CO_2, concentration of CO_2 and actual temperature rise), if the observation is different to expectations by a certain tolerance level then agents may change their world view. Four variables in the model are chosen to be uncertain and worldview dependant. The variables are: sensitivity of the temperature for increasing CO_2 concentration, the speed of technological improvement and energy conservation transition, mitigation costs, and damage costs. The management style is the actual decision making process which affects the economic environmental system. The type of management style implemented is determined by the weighted average of the individual perspectives.

The decisions made are based on the discrepancies between expected and observed outcomes of the system. The model is populated by 50 agents who compare the observed temperature change with the expected one. A fitness function is created which measures how well the system behavior (for a given management style) fits with actual observations. There is no empirical basis for such a fitness function, so in order to start the model one of the extreme world views is assumed to be the correct model of the system. The fitness function incorporates a tolerance level which represents how far the actual and expected temperature change have to be apart for action to take place, in other words it represents the agent's ignorance. While the results are described as tentative the authors conclude that the adaptive behavior of agents to climate change could make 'quite a difference'. The particular type of evolutionary economics employed by Janssen and De Vries has great potential to enhance our understanding of the social and economic consequences of climate change. The application of this type of analysis is complex, new and rapidly evolving and is currently outside the area of expertise of this book's author. Also, the nature of the method makes it highly unlikely at this stage that it could be applied successfully in an optimisation model such as SEADICE.

Each of the models mentioned fails to provide an optimisation treatment of the determination of the effects of adaptation on global warming. However, the bulk of the IAM literature does not provide any estimation of adaptation to climate change so simulation models such as those reviewed are useful at least in the respect that they have at least made an attempt. One of the important purposes of IAMs is to provide estimates of the impacts of climate change. It has already been established that for impact estimates to be seen as more reliable, the level of autonomous adaptation needs to be incorporated. Therefore, the experiments involving endogenous technical progress and autonomous adaptation in the following section provide an extension of the literature on IAMs of climate change that is a contribution to the literature.

5.7 An application to SEADICE

In this section an experiment is conducted on a version of the SEADICE model with endogenous technical progress, which is run to observe the possible effects of autonomous adaptation to climate change. As explained earlier, endogenous technical progress is considered an improvement upon how economic models might correctly represent autonomous climate change adaptation. It allows the interaction between climate change and economic growth to be modelled more realistically and therefore, go some way towards making these particular types of models more relevant for policy makers.

5.7.1 Application of endogenous technical progress to the SEADICE model

Following the approach adopted in Islam (1996, 2001), endogenous technical progress is introduced into the SEADICE model using the equation developed by Shell (1966), where the level of technical progress is dependant upon the amount of resources channelled to research and development (R&D) and the depreciation of technical knowledge. This equation represents a situation where the change in technology over time will be positively related to the resources allocated to knowledge creation (Maurseth 2001; Fedderke 2001). At the same time as older forms of technology become obsolete, knowledge is also subject to 'depreciation'. The equation can be represented as:

$$\frac{dA}{dt} = \alpha\sigma Y(t) - \beta A(t)$$

where A = the level of technology,
α = the fraction of output used for R&D,
σ = the research success coefficient
Y = output, and
β = the rate of decay of technological change.

Following from Islam (1996) the values assigned to the coefficients are 0.015, 0.04 and 0.285 respectively, based upon observed historical data.

Any introduction of R&D and innovation can play havoc with the mechanics of the standard neoclassical DICE type model. The subsequent potential for increasing returns, asymmetric information, oligopoly and sunk costs all have adverse consequences for the delicately balanced optimisation of a neoclassical economic growth model (Sheehan 2000). The difficulties associated with introducing endogenous technical progress into DICE type models is evident in Islam (1996) where several specifications of endogenous growth were attempted using the ADICE model. It was found that the only specification that would provide a stable solution was the Shell version of endogenous growth. Consequently, the Shell equation is used on the basis that this specification is most likely to provide a viable solution in a model such as SEADICE, which has the same basic structure as the ADICE model of Islam (1996).

In terms of representing climate change policy the endogenous version of the SEADICE model can be seen to represent a situation where a policy is undertaken to allocate more resources to knowledge creation and technical progress so that higher levels of autonomous adaptation are enabled. Therefore, if the value of α (the fraction of output used for R&D) is changed it can be said that this represents the impacts on the economy of autonomous adaptation. Figure 5.2 represents a situation where three different levels of a are assumed. These three situations can be described as low, middle and high levels of autonomous adaptation, where the values of a are 1%, 1.5% and 2% respectively. It can clearly be seen from Figure 5.2 that introducing these three rates of a into the model induces significant changes in economic output results. Therefore, the endogenous

Figure 5.2 Change in GDP at Different Levels of R&D

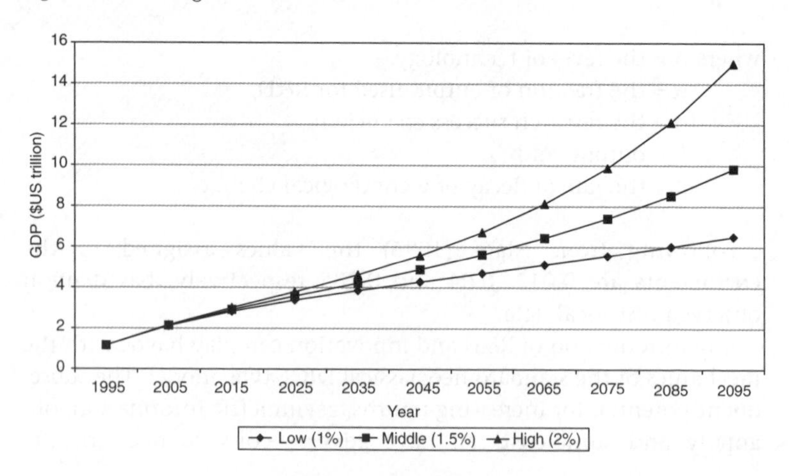

SEADICE model is sensitive to the resources devoted to knowledge and hence autonomous adaptation. If it is assumed this result represents reality then it supports the argument that autonomous adaptation is an important factor to consider as it can have significant effects on economic output. In this case if planned adaptation policies were based on impact estimates that did not incorporate autonomous adaptation the recommended policies could be incorrect as they may rely on economic forecasts that are significantly incorrect. If autonomous adaptation is as important as the SEADICE model suggests in these scenarios then it should become more of a priority subject for further research.

Throughout the literature pertaining to the economics of climate change significant uncertainties exist. All of the significant authors stress the fact that modelling attempts at representing any of the macroeconomic effects of climate change are to be considered illustrative at this early stage of the development of this field of economics (Cline 1992; Nordhaus 1994a; Fankhauser 1995b; Tol 1996; IPCC 2001b). This experiment to represent climate change adaptation shares an illustrative nature with the rest of the literature. However, this type of analysis should be regarded as a 'placeholder', which will be replaced by more accurate functional forms as our knowledge in this area of climate change economics improves (IPCC 2001b).

5.8 Conclusion

This chapter covers a great deal of ground in the area of adaptation to climate change. It began with a discussion of adaptation as a general concept where the general conclusion is that the concept can have many meanings in many different disciplines, which has the potential to lead to confusion when applying the concept in any intellectual framework. The focus is then narrowed to the discipline of economics, where the concept of adaptation has been somewhat of a problem. Only relatively recently has it been seriously considered through the branches of economics known as new growth theory and evolutionary economics. Attempts have been made by an increasing number of academics and institutions in recent times to define the concept of adaptation as applied to climate change economics. It is becoming more apparent in this literature that adaptation should be split into two types, autonomous and planned. Following this it is argued that planned adaptation is dependant upon the accurate representation of autonomous adaptation. However, throughout the relevant literature specifications of autonomous adaptation to climate change have not been attempted. It is assumed that the level of technology is an important determinant of the level of autonomous adaptation. This leads to the suggestion that a technique of new growth theory known as endogenous technical progress is a possible solution to the representation of autonomous adaptation. A review of IAMs reveals that only two of the many models produced in the last decade have attempted to incorporate climate change adaptation, and neither of those were optimisation models. Although applying endogenous growth theory to optimal growth models is difficult (Islam 1996), autonomous adaptation is modelled in the SEADICE model by using the Shell (1966) equation. The results indicate that the level of technological progress and therefore autonomous adaptation could be an important factor to consider for models of climate change impact. While the experiment in this chapter is purely theoretical and the results are illustrative it serves to highlight three important points. Firstly, that the difference between autonomous and planned adaptation is an important distinction to make, secondly that treating autonomous adaptation as dependant on the level of technology creates the potential for representing it in modelling frameworks and thirdly that economic growth models with endogenous technical progress are more likely to represent autonomous adaptation to climate change.

6
Mitigation Policy Options for South East Asia

6.1 Introduction

The objective of this chapter is to discuss the options for mitigation policy action in SEA. To achieve this the main mitigation policy options of the CDM, emissions trading, joint implementation (JI), carbon tax and no regrets are all explained and their relevance for the countries for SEA discussed. Given the characteristics of these options and the present international obligations, mitigation policy recommendations are made to conclude the chapter.

6.2 Policy options for greenhouse gas emission reductions in South East Asia

Currently none of the countries of SEA are in the UNFCCC decreed Annex I group. Therefore, none of the countries of SEA are currently under any obligation to reduce emissions as set forward in the Kyoto Protocol (Ghosh 2000). However, they can still be indirectly effected as a result of Annex I countries using the flexibility mechanisms available that make it possible to interact with non-Annex I countries. These flexibility mechanisms allow Annex I countries to meet their emission targets through means other than direct domestic mitigation. With the Kyoto Protocol now ratified, these flexibility mechanisms will become active components of international climate change. The international mitigation flexibility options of the CDM, JI and emissions trading will be reviewed in this section along with the options for domestic action such as a carbon tax or no regrets measures.

6.2.1 Clean development mechanism

Article 12 of the Kyoto Protocol provides the opportunity for Annex I countries to undertake projects with developing countries in order to reduce the emissions of the developing country in question and have that reduction count towards the target that the Annex I country has been set. This flexibility mechanism is called the CDM and is one of three flexibility mechanisms included in the Kyoto Protocol to enable multiple options for Annex I countries to meet their mitigation obligations. It allows for clean technology transfer to occur from developed countries to developing countries. The developing country potentially gains improved technology and higher foreign investment while the developed country gains Certified Emission Reduction (CER) credits that contribute towards Kyoto compliance, as well as any commercial profits arising from the investment. So far it is proposed that the project-by-project operation of the mechanism will be overseen by independent auditors known as Operational Entities (OEs). In the planning stages the project initiators will provide figures for both the environmental effectiveness (in terms of emission reduction) and commercial viability. Based on these estimates and the approval of all parties including the OEs a CDM project is then able to commence.

Even assuming that the Kyoto Protocol is ratified soon with no major changes there are several barriers that need to be addressed before CDM is a generally viable policy option (Ghosh 2000). Firstly, there are problems with the additionality criteria of the CDM. In Article 12 Paragraph 5(c) of the Kyoto Protocol it states that CERs can only be claimed if they are 'additional to any that would occur in the absence of certified project activity'. The economic justification of this criterion is that it can prevent the 'cherry picking' or freeriding of emission reduction projects that would have happened regardless of the Kyoto Protocol. However, this idea, while good in theory may have some problems in practice. Proving that a project would not have happened otherwise is a difficult task, even more so in a rapidly changing region such as SEA. This is another version of a problem which has been discussed in other contexts earlier in the book; that of the establishment of baseline data. From an economic viewpoint, where is the incentive for the private sector to invest in CDM activities that would not otherwise have occurred? It is not clear how the market for CERs will operate as yet. Will the private sector be able to sell CERs to their host country or any Annex I country? Or will the market be largely government to government trading? Will there be restrictions on the percentage of emissions reductions Annex I countries can claim as CERs? Will CDM projects be equally distributed among developing countries, or concentrated more on the most underdeveloped? While all of these questions are beyond the scope of this book and still to be worked out, the Asia Pacific Energy Research Centre (2001) identified several other factors specific to SEA that may also act as barriers to the CDM. These include: relatively higher risk premiums on capital; a general weakness of institutional frameworks; limited data for potential CDM investors; and uncertainty arising from possible instability of regulatory regimes and shifts in domestic energy policy. If work like this continues to identify specific barriers to CDM implementation in SEA then the framework could be provided to begin to minimise these barriers.

Most of the developing countries of SEA are cautious about many aspects of CDM but are nevertheless participating actively in the development process. For example, a meeting was held at the United Nations Industrial Development Organization (UNIDO) headquarters in Vienna on 27–29 August 2001, where representatives from Indonesia, Thailand, the Philippines, Malaysia and Vietnam implemented the initial phase of a program of activities focusing on industry, climate change and the CDM. Potential enabling institutions were identified along with the barriers to implementation of CDM in the region. The initial phase has already been completed on the way to the overall goal of providing a program with regional and national components that will facilitate CDM projects in the region (UNIDO 2001). This type of activity is encouraging, particularly as it is so early in the Kyoto process. If this is indicative of the way the policy makers of the region will attempt to embrace the CDM flexibility mechanism, the barriers that do exist may be overcome. Internationally, the COP7 meeting of November 2001 in Marrakesh saw the establishment of a 15 member Executive Board of the CDM which will begin implementation of the CDM in the near future. The main purpose of the CDM Executive Board is to enable the successful implementation of the CDM by eliminating the barriers mentioned earlier in this section.[31] CDM is potentially the most viable and effective regional policy option for SEA as long as the above issues related to the implementation of the CDM are resolved to ensure regional interests and development perspectives.

6.2.2 Emissions trading

Under Article 6 of the Kyoto Protocol, reference is made to the provision for emissions trading. However, it is worded in such a way that the intended outcome is vague. The mitigation policy option of a tradable carbon emission permit scheme has been examined by many authors, even before the Kyoto Protocol (Nordhaus 1994a; Fankhauser 1995b). The proposed policy consists of the creation of a fixed level of carbon permits that allow the holders to emit the pre-determined amount of carbon that their permit allows. These permits should be made tradeable to allow the most efficient allocation of carbon emissions amongst the trading parties. The policy should be successful as long as the marginal cost of reducing CO_2 emissions is different amongst trading partners. If this is the case the permit holders have an incentive to trade permits where those parties with the higher marginal mitigation costs are prepared to buy permits from those with lower mitigation costs. The process would continue until marginal mitigation costs are equalised across parties and hence a cost-efficient distribution of CO_2 emissions would be achieved. The bulk of emission cutbacks would be distributed to those parties most able to afford mitigation.

However, many considerations must be addressed before such a policy would be viable on a regional or global scale. Factors such as the measurement of emissions, initial setting of emission limits, and initial allocation of permits, are a few among many obstacles facing the implementation of a tradeable carbon emissions policy. As yet, the author is unaware of any steps that have been made by governments in SEA towards implementing such a policy. The developing countries are also wary of the Annex I countries using the process as a way of avoiding the cutting of domestic emissions by buying cheap credits. The administrative costs of implementing such a scheme may be so large that marginal costs may be distorted to such an extent that trading is impractical in some or all cases. Disagreement also exists between many domestic science communities with the emission assessments conducted by the international science community. Kandlikar and Sagar (1997) give the example of methane emissions in India where very large discrepancies exist between the United States Environmental Protection Agency who estimated rice paddy methane emissions of 37.8 Tg/year, and the Indian Methane Campaign where 4 Tg/year was estimated. Kandlikar and Sagar (1997) claim that these findings (which take into account local agricultural methods and soil conditions) have been largely ignored in the international literature, and that scientists from the developed countries may be disconnected from the particular needs, realities and interests of the developing countries.

6.2.3 Joint implementation

JI is another Kyoto flexibility mechanism where an Annex I country invests in emissions mitigation project(s) in other Annex I countries where the costs of mitigation are lower than domestically and the country is credited with the mitigation from its own emission total. The most important aspect of JI for the purposes of this book is that it can only be implemented between Annex I countries. Therefore, this means that JI is not a policy option for any of the countries of SEA.

6.2.4 No regrets mitigation options for South East Asia

No regrets measures are those for which benefits, such as reduced energy costs and reduced emissions of local pollutants equal or exceed the cost to society, excluding the benefits of climate change mitigation (Cline 1992; Fankhauser 1995b; IPCC 2001c). These actions are sometimes known as measures worth doing anyway. The possibility for no regrets measures exists because market failures and distortions are present. Market failures and distortions such as imperfect information, subsidies, etc. occur when incentives offered to individuals, households and firms encourage behavior that does not meet efficiency criteria (i.e. private and social prices diverge) and can prevent otherwise profitable emission reduction investment taking place. If these market failures can be clearly identified then any country or region that makes climate change a priority can exploit their no regrets potential by enabling otherwise economically and environmentally efficient practices. However this is also

Table 6.1 Sectors Where Dual Environmental Benefits are Possible

	Domestic Benefits	Climate Change Benefits
Forestry	• Biodiversity • Existence value	• Sequestered CO_2
Energy Use	• Health • Energy conservation • Reduced air pollution	• Reduced emissions
Agriculture	• Increased volume • Decreased variability	• Reduced emissions
Population	• Reduced stress on resources and infrastructure	• Reduced emissions

not an option without problems, identifying no regrets measures does not come at zero cost, they can be difficult to identify and the outcomes may be uncertain. Therefore, until policy makers are confident enough in the potential for efficiency improvements, no regrets measures will not be undertaken.

Few extensive studies have been done to identify mitigation policy options in SEA, even fewer have been focused on no regrets policies. The outstanding contribution in this area to date has been ALGAS. This comprehensive study covered 12 Asian countries and produced emissions inventories and mitigation strategies for each (ALGAS 1998a, 1998b, 1998c, 1998d). The following sections review some of the mitigation policy options that have been suggested for two of the most important sectors throughout SEA. Table 6.1 illustrates in a simple form some of the possible domestic environmental issues that can be addressed that have ancillary climate change benefits.

6.2.4.1 Mitigation policy options for land use in South East Asia

Land use (and the accompanying issues of forestry management and urban development) is an important sector throughout SEA both for economic reasons and also for climate change (Bautista 1990). As mentioned in Section 2.3.1 of this book; deforestation is a major contributor to SEA's climate change emissions and is a more important factor relative to world standards. Land use issues have a high profile at the community and government level as an environmental issue in SEA, and have accounted for serious environmental problems apart from climate change emissions such as lower water tables, flooding, topsoil loss and reduction in biodiversity (Asian Development Bank 2000).

In terms of no regrets policy options, market failures do exist in the SEA forestry/land use sector. Primarily they result from either misallocated or unrecognised property rights issues or poor government policy with unintended negative consequences. In Indonesia there is a typical case of how incorrect policy measures can lead to market imperfections and unsustainable levels of deforestation. Trade restrictions imposed on the export of unprocessed log and wood products led to an oversupply in the domestic market where prices were about half that of the world level. Higher fees, if placed on the value of the standing timber rather than the processed wood would internalise the environmental costs of deforestation and encourage more efficient logging practices (Brandon and Ramankutty 1993). Another example occurred in the Philippines where macroeconomic policies that subsidised the exports of manufactured goods and that taxed labour at a higher rate relative to capital compared to developed countries led to the displacement of many labourers and their families, who migrated to upland public forests. Once there they cleared parcels of forest to grow crops, which led to soil erosion, downstream sedimentation of reservoirs, harmed offshore coral reefs and fisheries and depleted soil fertility (Habito 1993). These examples demonstrate that macroeconomic policies can have consequences, which lead to market failures for sectors such as land use.

The level of emission savings which are possible from reduced deforestation and improved land use in SEA are still uncertain. However, some work has been done in this area. An example is the 1999 country study of Vietnam by the Hydrometeorological Service of Vietnam where mitigation options for that country were examined (Hydrometeorological Service of Vietnam 1999). It was found that the mitigation options of enhanced natural regeneration, reforestation, natural forest protection and scattered trees are all viable policy options with positive net present value results. It was also found that up to 5,500 Tg of carbon could be mitigated for Vietnam without economic cost. In Malaysia there is a 25 year project underway between a logging concession holder and the Forest Absorbing

Carbon Dioxide Emissions Foundation to rehabilitate 25,000Ha of degraded logged forest to sequester 5 Mt of carbon at a total cost of $US14 million. This roughly equates the cost to around $US3 per ton of carbon (IPCC 1996b). This demonstrates that potentially significant low or no cost emission reduction potential exists in this sector for SEA. Perhaps this is incentive enough for the countries of the region to prioritise no regrets mitigation policy for this sector and allocate resources for realising this potential.

6.2.4.2 Policy options for the energy sector

The energy sector is potentially the most important source for the mitigation of GHG emissions (Malik 1994). The demand for energy in Asia is doubling every 12 years, much faster than the world average of 28 years (Brandon and Ramankutty 1993). While the very high growth of demand for energy may be detrimental for the continued emissions of GHGs in one sense, it has a positive side in that high growth in this sector provides opportunities for achieving increasing efficiency. The rapid growth of energy production and demand will provide the potential opportunities for bypassing or removing market failures by investing in or promoting leading edge technologies. Market failures commonly take on two main forms in the energy sector throughout SEA (Sharma 1994). Firstly, energy prices in SEA are often highly subsidised, and therefore industries have a reduced exposure to competitive pressure to reduce costs or introduce new products which lead to environmental and social externalities. Secondly, the availability and flow of information is restricted in the sense that norms and regulations are neither as transparent or widely enforced as in developed economies. As a result, price reform and information programs may not stimulate sufficient improvements and investments in energy efficiency.

It is precisely the second market failure mentioned that the countries of SEA are focusing on for no regrets mitigation. In their National Communications to the UNFCCC the countries of Thailand, Indonesia, Malaysia, Singapore and the Philippines, all expressed the desire to approach the exploitation of no regrets mitigation in the energy sector initially through DSM. DSM includes options for reducing the demand for energy by the introduction of more efficient energy use either through legislation, direct intervention or public awareness campaigns. CO_2 emissions per unit of GNP in developing SEA countries are significantly higher than in industrialised nations.

Potentially this means that significant inefficiencies are present in the energy sector in SEA such as outdated methods or aging capital. If these inefficiencies can be identified through DSM then demand for energy and therefore emissions can be reduced. This has already been done in Thailand where the relevant authorities implemented a five year DSM program in 1992 (Office of Environmental Policy and Planning 2000). This program was very successful and achieved energy reductions over double that expected. Currently a more ambitious DSM is under way and is already reducing energy use through measures such as replacement of energy efficient fluorescent lamps and the promotion of energy efficient electrical appliances. The ALGAS (1998c) study on Thailand identified DSM dealing with residential and commercial lighting, commercial cooling and refrigerators which all displayed no regret properties. It is therefore apparent that DSM is not only the most desired policy from within SEA but also it has already been indicated that if the appropriate research can be conducted then no regrets energy mitigation policy is possible and in the case of Thailand is already taking place.

6.2.5 Carbon tax

A carbon tax is simply a tax applied on a per unit basis (i.e. per ton) of the emissions of carbon made by firms. The purpose of a carbon tax is to encourage the reduction of emissions by making activities that produce carbon emissions more expensive and therefore, lower emission alternatives relatively cheaper. Proposals for a carbon tax have existed for quite some time and many studies have attempted to calculate the level of tax that would be required to stabilise GHG concentrations (Cline 1992; Nordhaus 1994a). The policy may have some practical difficulties such as the method of implementation, measurement of emissions, setting the correct tax rate, migration of carbon intensive industries and also the pressure from lobby groups on government. There are other issues such as; who receives the revenue, what purposes should the revenue be used for, what level should the tax be, how are emissions to be measured and what emissions should be included? It has been estimated that in order to stabilise CO_2 emissions for the developing Asian countries a carbon tax would have to be set at $US500 per ton (in 1985 prices) in the year 2050 (Tomitate 1991). If levels of this magnitude are needed carbon taxes may not be viable, as they could be a large constraint on economic development.

As an experiment the SEADICE model was run to determine the optimal carbon tax for SEA. In SEADICE the carbon tax is determined as a Pareto-optimal policy which induces the economically efficient level of emissions that balances current mitigation costs against the future environmental benefits of carbon mitigation. This can be attained by setting the carbon tax equal to the global environmental shadow price of carbon. The environmental shadow price of carbon is the effect by environmental means of a unit of emissions today on the present value of consumption in all future periods. In the experiment with the SEADICE model a cooperative global regime is assumed where the ROW emissions follow a trajectory equivalent to achieving $2\times CO_2$ concentrations. Therefore, the results presented here represent the optimal carbon tax policy given this scenario. What the results reveal is that an optimal carbon tax would yield at global $2\times CO_2$ benchmark a carbon tax of $US31.40, emission reduction of 7% and an impact on GDP of –0.02%. While these results are only illustrative,[32] they reveal that given the assumptions of the SEADICE model the effects of an optimal carbon tax across SEA would be minimal. This means that in terms of emission reductions the effectiveness of an optimal carbon tax is quite moderate and therefore diminishes its potential for future implementation. This book does not attempt to find carbon tax rates by implementing SEADICE model runs such as climate stabilisation targets for the reason that the total emissions from SEA are so small that any attempt at modelling such policy options are futile. To explain, SEA emissions make up around 2% of global emissions. In the SEADICE model, if the model is run for a scenario where concentrations need to be kept below a certain level, say $2\times CO_2$, the model can only control the emissions from SEA which are only 2% of global emissions. If the $2\times CO_2$ scenario requires a global emission reduction of more than 2% then the model cannot be solved. Models based on the original globally aggregated DICE model are suited to finding optimal climate stabilisation emission paths.

6.3 Expected climate change mitigation actions for South East Asia

Given the mitigation options reviewed earlier in this chapter, what are the expected future mitigation actions in SEA? In June 1992, all of the countries of SEA except Cambodia, Malaysia and Laos signed the UNFCCC. By August of 1997 all of the countries of SEA including Cambodia, Malaysia and Laos had ratified the convention.[33] The objective of the convention is to stabilise atmospheric concentrations of GHG, however currently the countries of SEA are not legally bound to emission targets specified by the FCCC and Kyoto Protocol as they fall outside of the list of Annex I countries. This situation is not likely to change until at least 2008–12, which is the first target date for Annex I emission reduction goals. Therefore, significant mitigation policies are not expected from within SEA for at least a decade, although policies focused on climate change, while not significant at present are a specific focus for most of the environmental government departments of the region.

6.4 Recommended mitigation actions for South East Asia

Based on the fact that no emission reduction will be enforced on any SEA country over the next decade two emission mitigation avenues are recommended for the countries of SEA to pursue over the next decade. Firstly, of the flexibility mechanisms in the Kyoto Protocol, CDM is the only one likely to provide opportunities for SEA. The CDM has the potential to provide additional foreign direct investment with consequential technology transfer and the economic benefits which they can deliver. Of course this all depends on the extent to which the CDM is used by the developed countries. If there is little or no demand for the flexibility mechanism the countries of SEA will be competing for limited opportunities. Secondly, the existence of barriers to no regrets mitigation measures should be identified and removed, this chapter has revealed that the focus should be on the forestry and energy sectors as significant potential exists for the exploitation of no regrets mitigation measures in each of these sectors. The emphasis for government at this stage should be on the possible gains in economic efficiency that are available by enabling no regrets mitigation measures. At the moment the measurement of the potential environmental benefits of such policies is highly uncertain, therefore a focus on the more easily estimated economic benefits of removing market imperfections in the forestry and energy sectors are much more likely to be implemented. As revealed in this chapter several SEA countries have expressed a preference to exploit no regrets mitigation in the energy sector through DSM. Indeed, Thailand has already implemented one DSM program and is currently part of the way through another. The success of these programs and the willingness of the region to adopt DSM techniques are encouraging for the likelihood of GHG mitigation in the region. Both of the options mentioned in this section are also important when considering the current opportunities available in SEA with respect to the potential for alternative development paths of climate change mitigation technology (Forsyth 1999; Angel and Rock 2000).

6.5 Conclusion

This chapter is not aimed at being a comprehensive treatment of mitigation options for SEA. It is a topic that has been covered comprehensively already in many studies (Bhattacharya, Pittock and Lucas 1994; Malik 1994; Qureshi and Hobbie 1994; Sharma 1994; ALGAS 1998a; 1998b; 1998c; 1998d). The main focus of the book is the impacts and adaptation to climate change for SEA. In keeping with maintaining this focus this chapter has given a brief overview of the mitigation options available to SEA. Given the policies available and their characteristics, the policy options of the pursuit of CDM projects and focused DSM no regrets mitigation alternatives are recommended. A topic that has received far less attention though is the adaptation policy options available for SEA and is the focus of the next chapter.

7
Adaptation Policy Options for South East Asia

7.1 Introduction

If the predictions of the IPCC (2001a) are correct then the rates of climate change will test many of the limits of human adaptation in the decades to come. As Yohe (2000, p. 371) states, 'the research community has long passed the point of considering adaptation in the abstract'. Where Chapter 5 used positive analysis to determine what adaptations are likely, this chapter will be a normative exercise where the key question is: What adaptations are recommended?[34] The chapter begins by defining climate change adaptation policy and then examining its treatment in the international context and why its stature is increasing. This is followed by a discussion of the range of methods that have been used to identify possible adaptation policy options, and whether they are likely to be implemented. Given the significance of adaptation policy, the options that are available, the modelling results and relevant discussions from previous chapters, and based on these factors recommendations are made regarding the future path for the identification of adaptation policies for SEA.

7.2 Definition of climate change adaptation policy

7.2.1 What is adaptation policy?

Adaptation policy is any strategic action taken which results in the reduction of vulnerability to the effects of climate change (Benioff, Guill and Lee 1996; Fankhauser 1996; Smith 1997; Pielke 1998;

UNEP Collaborating Centre on Energy and the Environment 1998). Given that policy formulation by definition involves some form of planning, adaptation policies will be conducted in anticipation of climate change as well as in reaction to climate change. Five generic objectives of adaptation policy have been identified by Klein and Tol (1997):

1. To increase the robustness of long term investments and infrastructure.
2. Enhancing the flexibility of vulnerable managed systems.
3. Improving the adaptability of vulnerable natural systems.
4. Reversing current cases of maladaptation which are increasing vulnerability.
5. Increasing societal awareness and preparedness for climate change.

7.2.2 What is adaptation's status as an international policy issue?

If adaptation is to be taken seriously as a policy issue by regions such as SEA then it must be recognised within the most important climate change institutions. There are several references to adaptation policy in the FCCC which are summarised as follows:

> 'The Parties should take precautionary measures ... To achieve this, such policies and measures should take into account different socio-economic contexts, be comprehensive, cover all relevant sources, sinks and reservoirs of greenhouse gases and adaptation ...' (Article 3, Section 3).
>
> 'All parties .... shall ...(Article 4);
>
> Formulate, implement, publish and regularly update ... measures to facilitate adequate adaptation to climate change (Article 4, Section 1(b));
>
> Cooperate in preparing for adaptation to the impacts of climate change (Article 4, Section 1(e));
>
> Take climate change considerations into account, to the extent feasible, in their relevant social, economic and environmental policies and actions, and employ appropriate methods, for example impact assessments, formulated and determined nationally, with a view to minimizing adverse effects on the economy, on public health and on the quality of the environment, or pro-

jects or measures undertaken by them to mitigate or adapt to climate change (Article 4, Section 1(f)) and

The developed country Parties and other developed Parties included in Annex II shall also assist the developing country Parties that are particularly vulnerable to the adverse effects of climate change in meeting costs of adaptation to those adverse effects.' (Article 4, Section 4).

The five clauses within the UNFCCC referred to above that pertain to climate change adaptation refer to the need for all parties to the protocol to implement measures to facilitate adaptation, to cooperate in doing so, to minimise the impact of implementing adaptation policies and for the developed country parties to assist developing country parties. Compared to the proportion of the document devoted to mitigation, these references to adaptation are minimal. This reflects the historical importance that has been placed upon the two policy options by the international community, an issue that was discussed in Chapter 1. These references to adaptation are quite broad and non-specific, however this is all changing with the introduction of the Adaptation Policy Framework (APF). The APF has been created from the IPCC/UNFCCC process and aims to develop national planning and development strategies for climate change adaptation. These strategies are intended to facilitate the identification and implementation of climate change adaptation policies in developing countries. At the moment the initial stages of this program are nearing completion. Based on a review of the draft literature thus far the author anticipates that the APF will be the most important step yet on the road to realistic adaptation policy development for developing nations. Other action towards climate change policy is occurring in the SEA region as is highlighted by the recent Thematic Workshop on Vulnerability and Adaptation Assessment which was held on 10–12 May 2000 in Jakarta (Page 2001). As an international policy issue adaptation is becoming increasingly important as the final details of the international mitigation policy framework are decided.

7.2.3 Reasons for the increasing importance of climate change adaptation policy

The consideration of adaptation as a climate change policy option has been overshadowed by mitigation, however some argue that this imbalance should be redressed (Pielke 1998). This balance will be influenced by several factors that either reveal that adaptation should be more important, or over time will force more attention on the issue. In this section these factors are explained.

An important characteristic of adaptation is that it has direct local benefits, whereas mitigation policies will result in global benefits. For example the installation of an irrigation system would have benefits such as a reduction of agriculture's vulnerability to climate change. However, unlike mitigation the benefits accrue to the area that has been covered by the irrigation. The characteristic of direct local benefits means that adaptation policy is less vulnerable to free-riding behavior and is also a desirable factor for policy makers.

Certain adaptation policies are also characterised for having immediate benefits. If a sea wall is built its effects are immediate whereas any benefits obtained from mitigation would not be realised for many decades. This is an important consideration for policy makers, as quick results make implementation easier.

As Ausubel (1993) notes, vulnerability to the effects of climate change are decreasing as advances in technological and social developments occur. This is true for developed nations in general but has not been conclusively determined for developing countries. Facts that have been used to substantiate this argument include the reduced incidence of deaths from tornadoes in the United States over the period 1917–90. A decrease in the death rate from natural disasters has been seen across many developed nations as technology and techniques reduce societies' vulnerability to the extremes of climate (Albala-Bertrand 1993; IFRCRCS 1997; Bruce 1999). This implies that adaptation may become less important over time if our vulnerability to climate change substantially decreases over time. However, it is possible that some countries or regions are experiencing increasing vulnerability to climate change. If this is the case, adaptation policies will become very important once the problem of increasing vulnerability is identified more clearly (Kelly and Adger 1997).

Levels of vulnerability to climate change may be so high that it might already be in the best interests of certain countries to pursue adaptation as a higher priority than mitigation. Many of the small island states have emission levels so small relative to global levels that any attempts at emission reduction are largely futile and would be mostly symbolic of their commitment to combating climate change. The number of countries that may be in this category is unknown at this stage.

Some climate change is already inevitable; therefore adaptation eventually will be required (Fankhauser 1995b). Even if all emissions of GHG were halted today some warming will eventuate as a result of the lagged effects of past emissions (the warming effects of current emissions will not occur for many years). Nordhaus (1994a) estimated that if a policy to cap emissions at 80% of 1990 levels were adopted (which would require emission reductions of 70% late into next century) global temperatures would still rise by 2.2°C by 2100. Given the current level of mitigation targets set out in the Kyoto Protocol this type of scenario would require major technological breakthroughs and peaceful worldwide policy regimes into the next century. If this does not eventuate then the situation will be one where 'mitigate we might; adapt we must' (Nordhaus 1994a, p. 189). Authors such as Cline (1992) have already warned that the problem is in essence a permanent one as CO_2 concentrations will linger for hundreds of years therefore we must be prepared to combat the likely climate change scenarios of beyond $2{\times}CO_2$.

While the importance of adaptation options will not and should not exceed that of mitigation in the near future, adaptation policies will become more viable as it becomes more obvious that concentrations of GHG will keep on rising.

7.2.4 Options for the identification of adaptation policies

What are the economic methods that have been used to identify and assess climate change adaptation policies? Much of the following material has been adapted from two main sources, Feenstra et al. (1998) and Stratus Consulting (1999). 'In general, an approach to estimate (in either qualitative or quantitative terms) both the costs of implementing a measure and the potential benefits from doing so is needed' (Feenstra et al. 1998, s. 5–11).

7.2.4.1 Generic procedures for adaptation policy assessments

Both the IPCC and United States Country Studies Program (USCSP) have provided very general procedures for adaptation policy assessments.[35] They have been made deliberately general so that those who follow the procedure have a substantial scope to choose a particular methodology to implement the assessment. The main reason for this has been to allow flexibility for policy makers as many of the methodologies are still being developed and may be specific to particular regions or sectors.

The IPCC method. While it is not within the IPCC charter to provide specific policy recommendations, it has provided a procedure for policy makers to identify adaptation policy options and therefore assist countries in meeting commitments under Article 4 of the UNFCCC. Although this procedure is provided in the IPCC (1996b) volume dedicated to scientific assessments the following procedure could be adapted for economic assessments as well. The IPCC (1996b) recommends a seven-step procedure for a climate impact and adaptations assessment:

1) Definition of the problem.
   This step requires identifying the goals of the assessment, the physical, environmental and economic range of the assessment, and the data needs including time horizons.
2) Selection of the method.
   This step is dependent on the availability of data, skills, models, resources and other factors that will influence the methodology of an adaptation assessment.
3) Testing the method.
   Model validation and sensitivity studies should be completed to ensure the robustness of the methodology to be used. Three types of testing are suggested including: feasibility studies, data acquisition and compilation and model testing.
4) Selection of scenarios.
   This requires monitoring existing climate conditions and using those baselines to extrapolate future climate change scenarios.
5) Assessment of biophysical and socioeconomic impacts.
6) Assessment of autonomous adjustments.
7) Evaluation of adaptation strategies.
   a) Define the objectives.
      i) Goals need to be identified such as sustainable development or the reduction of vulnerability.
   b) Specify the climate impacts of importance.
   c) Identify the adaptation options.
   d) Examine the constraints.

e) Quantify measures and formulate alternative strategies.
f) Weigh objectives and evaluate tradeoffs.
g) Recommend adaptation measures.

This procedure is given in more detail in a book entitled 'IPCC Technical Guidelines for Assessing Climate Change Impacts and Adaptations' (Carter et al. 1994).

The USCSP method. The USCSP provided financial and technical assistance to 56 developing countries in the mid 1990s to develop emissions inventories and also to evaluate mitigation and adaptation strategies (Smith et al. 1996). The process for identifying and implementing climate change adaptation policy used by the USCSP constitutes six steps as outlined in Benioff, Guill and Lee (1996). The steps in point form are:

1) Define the scope of the problem(s) and assessment process.
2) Choose scenarios.
3) Conduct biophysical and economic impact assessments and evaluate adaptive adjustments.
4) Integrate impact results.
5) Analyse adaptation policies and programs.
6) Document and present results.

Other methodologies are also available to identify adaptation policy options as can be seen in Table 7.1. The methods shown in this table range from the qualitatively simple to the quantitatively difficult. For a more detailed analysis of the many options for adaptation policy identification see UNEP Collaborating Centre on Energy and the Environment (1998).

7.2.4.2 Results of adaptation policy identification

The range of possible policy options that can be used for adapting to climate change is substantial. This is illustrated by Table 7.2 which highlights many of the policy options identified by Benioff, Guill and Lee (1996).

7.2.5 Difficulties of adaptation policy identification

With regard to the barriers that exist for adaptation policy identification, one in particular stands out. In order to measure the

Table 7.1 Adaptation Assessment Methodologies

Adaptation Assessment Methodology	Description
Forecasting by Analogy	Qualitative method where comparisons are made of observed historical adaptations that were made in climate conditions similar to likely future climate change conditions.
Expert Judgement	The opinions of individuals with particular expertise are used to glean information pertaining to adaptation policy options. Mostly used in a panel format where the individual results are aggregated to elicit a broader opinion.
TEAM Software	Software developed by Decision Focus (1996) for the United States Environment Protection Agency (USEPA). Qualitative and quantitative criteria are used in this software package to compare adaptation strategies.
Adaptation Decision Matrix	This method is useful when the majority of benefits obtained from achieving policy objectives cannot be monetised or expressed in a common metric. To avoid subjectivity, detailed analysis is needed in order to provide a substantial basis for the matrix. This method is presented in Smith et al. (1996).
Cost Benefit Analysis	Used to determine whether an adaptation policy is economically justified. This method generally involves two steps: identify and screen the benefits and costs to be included; and then convert them into monetary values where possible.
Cost Effectiveness Analysis	Applicable when it is difficult to quantify and monetise benefits. Adaptation measures are compared by determining their cost differences for achieving a fixed level of benefit.
Implementation Analysis	The goal is to find the least costly policy measure in terms of relevant factors such as money, time, etc. Applicable when the assumption is made that the benefits of different adaptation measures are comparable. The implementation barriers should be identified along with the difficulty of each, this can be done using a matrix to enable decision making.

Table 7.2 Typical Adaptation Policies

Category	Adaptation Policy Option
General	• Assess current practices of crisis management • Inventory of existing practice and decisions used to adapt to different climates • Promote awareness of climatic variability and change
Agriculture	• Develop new crop types and enhance seed banks • Liberalise agricultural trade • Avoid tying subsidies or taxes to type of crop and acreage • Promote agricultural drought management • Increase efficiency of irrigation • Disperse information on conservation management practices
Forests	• Encourage diverse management practices • Reduce habitat fragmentation and promote development of migration corridors • Enhance forest seed banks • Establish flexible criteria for intervention
Water Resources	• Use river basin planning and coordination • Adopt contingency planning for drought • Make marginal changes in construction of storage and distribution facilities • Maintain options to develop new dam sites • Conserve water • Allocate water supplies by using market-based systems • Use interbasin transfers • Control pollution
Sea Level Rise	• Adopt coastal zone management • Use presumed mobility • Plan urban growth • Discourage permanent shoreline stabilisation • Incorporate increases in the height of coastal infrastructure • Preserve vulnerable wetlands • Decrease subsidies to sensitive lands • Tie disaster relief to hazard-reduction programs • Promote public education
Ecosystems	• Integrate ecosystem planning and management • Protect and enhance migration corridors or buffer zones • Enhance methods to protect biodiversity off-site

Source: Benioff, Guill and Lee (1996, Section 7-5).

benefit side of adaptation policy options the economic impact of particular climate change measures must be estimated. To estimate the economic impact of climate change the vulnerability of the economy to climate change must be known. It was demonstrated in Chapters 3 and 5 that this procedure is very difficult and characterised by high uncertainty. For the individual countries of SEA the difficulties associated with this type of analyses may be too great because of a lack of resources or expertise. Given the characteristics discussed thus far it is now possible to recommend the appropriate approach to adaptation policy for SEA.

7.3 Adaptation policy options for South East Asia

Using modelling results and the arguments and discussions from both the current and previous chapters of this book, recommendations are made for SEA with respect to adaptation policy in this section.

The modelling results from Chapter 4 forecast negative economic impacts across SEA for $2\times CO_2$ climate change conditions in the order of 5% of GDP. If this is the likely extent of climate change impact on SEA then significant scope exists for the pursuit of adaptation policy options. The reason this allows for more adaptation options is that any reduction of the 5% negative impact can be considered a direct benefit to the economy. Any adaptation policy that can be identified that reduces the 5% negative impact by more than it costs to implement the policy is economically viable. In other words the higher the vulnerability of an economy to climate change the greater is the scope for adaptation policies to be adopted. As this book has identified, SEA is a region with relatively high vulnerability, therefore the options for adaptation policy are greater than other regions. The modelling results of Chapter 5 also highlight the importance of adaptation as a policy option for SEA. In Chapter 5 the experiment of including endogenous technical progress in the SEADICE model indicated that the level of autonomous adaptation could have a significant effect on economic outcomes for SEA. Therefore, any policies that can be identified that enable increased levels of autonomous adaptation could reduce climate change vulnerability in the region. Given the modelling results of this book it is recommended that SEA should prioritise adaptation policies as the initial focus of climate change policy identification. If adaptation policies are to be pursued in SEA, then how will they be identified? An answer to this question is undertaken in the following section.

The identification of adaptation policies for SEA is likely to become increasingly important as more is understood about the region's vulnerabilities to climate change. Several factors have been identified in this book that will influence the recommendations in this section. Firstly, the modelling results of Chapter 4 revealed that SEA is more vulnerable to the effects of climate change than developed regions, thus making adaptation relatively more important. In Section 7.2.5 of this chapter the argument was made that while many methods do exist to identify policy options, they rely on the estimation of the impacts of climate change which are difficult to establish. Following the discussions of potential climate change impacts on the region in Chapter 3 it was concluded that several broad vulnerabilities are shared throughout the region, these being; (1) long vulnerable coastlines, (2) similar tropical ecosystems, (3) similar agricultural outputs and vulnerabilities, (4) shared concerns arising from rapid economic development. These shared similarities provide synergies that can facilitate regional cooperation in the identification of no regrets adaptation policies.

All of these factors support the argument for the pooling of resources with respect to the identification of climate change adaptation policies for SEA. To summarise, pooling resources makes sense in this case because:

a) SEA is more vulnerable to the effects of climate change than developed regions.
b) Policy identification is difficult and resource intensive.
c) Many specific climate change vulnerabilities are shared throughout the region.

In order to cooperate on a regional level to identify adaptation policy options a regional institution of some type is required. A list of regional environmental institutions that are either multilateral or have a regional philosophy in SEA is provided in Table 7.3 as a starting point for this discussion.

The formation of environmental institutions concerned with the issues of SEA has expanded dramatically since the 1992 Rio Earth

Table 7.3 Multilateral Environmental Organisations in South East Asia

- United Nations Economic and Social Commission for Asia and the Pacific
- ASEAN Environment Program
- Global Environment Facility
- World Bank
- Asian Development Bank
- Southeast Asian Nations Environmental Programme
- Mekong River Commission
- Regional Network of Research and Training Centres on Desertification Control in Asia
- Forestry Research Support Programme for Asia and the Pacific
- Asia Least-Cost Greenhouse Gas Abatement Strategy
- Coastal and Marine Environment Management Information System
- Southeast Asian Regional Committee for START
- Asia Pacific Network for Global Change Research
- Global Change Impacts Centre for Southeast Asia

Summit. Table 7.3, while not being comprehensive, provides an illustration of the extent of multilateral and regionally focused environmental organisations working in SEA. The large number of institutions has led to duplication of resources, which has been described as 'overwhelming' by some (Rogers 1993). Authors such as Amadore et al. (1996) and Jalal (1993), along with many others, agree on the general principle that the nations of SEA should participate in effective multilateral cooperation on climate change policies, although others such as Drysdale and Huang (1995) argue that there could be difficulties with regional cooperation, because countries in the region are at different stages of development. Cooperation is important in order to share in knowledge and skills and while difficulties may arise they should not be significant enough to abandon something as significant as climate change policy formulation. It is apparent from the multitude of institutions and the possibilities for duplication that the creation of any new multilateral environmental institution in SEA is not warranted in the near future. Therefore, it can be concluded that an existing institution would be the logical choice to focus the region's demand for adaptation policy.

ASEAN is one of the most influential regional institutions in SEA at this time, although its primary focus is not environmental. However, the ASEAN ministers have already met and passed several environmental resolutions; the 1990 Kuala Lumpur Accord on

Environment and Development, the 1992 Singapore Resolution on the Environment, the 1994 Bandar Seri Begawan Resolution on Environment and Development, the 1995 ASEAN Cooperation Plan on Transboundary Pollution and the 2000 Kota Kinabalu Resolution on the Environment. Therefore, environmental expertise and discourse already exists within the institution. Over the years ASEAN has transformed. As Rosenberg (1999) notes, from 1997 the previous ASEAN policy of non-interference and non-confrontation has begun to erode. The fire haze disasters that periodically caused enormous disruption in the region through the 1990s were the catalyst for the first major regional environmental collaboration. It created an early warning system that involves satellite data from Singapore, fire prevention from Malaysia and fire-fighting from Indonesia. The ASEAN ministers have been meeting monthly since 1997 in an attempt to keep on top of the problem. As Rosenberg (1999) observes, this appears to be a genuine attempt at dealing with a transboundary environmental problem. This is an important precedent for environmental problems such as climate change. The same sort of cooperation could be applied to the problem of adaptation to climate change, most likely through the ASEAN Environment Program (ASEP). The ASEP, which was implemented in 1978, is the vehicle by which ASEAN promotes sustainable economic development by the proper management of the environment.[36]

There is support for this argument at the government level within SEA as well. In Thailand's National Communication to the UNFCCC it is stated that:

> Thailand supports the use of regional cooperation as a means of sharing information and experiences on climate change issues. At the sub-regional level, Thailand views ASEAN as an important forum for offering support for implementation of the climate change convention. Sub-regional cooperation could focus on cooperation in research and development on climate change issues. The similarity of cultures and economic structures among ASEAN member countries could enhance the application of models to the sub-region. ... The exchange of information and experiences will accelerate the capacity building process in the sub-region and the region. (Office of Environmental Policy and Planning 2000, p. 78)

Therefore, it is recommended that ASEAN be considered as the most appropriate institution to carry out climate change adaptation policy identification for SEA.[37] In summary, it has been recommended for the following reasons: its influence and access to resources throughout the region; its recent shift to interventionist actions in the region and its record of environmental awareness highlighted by the recent cooperative policy action on smoke haze. With the initial stages of the APF due soon the methodological base may become available to further guide adaptation policy identification for the region.

7.4 Conclusion

Adaptation policy has been, and probably always will be the junior partner to mitigation policy. However, its profile is gradually expanding with the international climate change research community, especially with the recent establishment of the APF. Given the modelling results of Chapter 4, which indicated that SEA is relatively highly vulnerable to the impacts of climate change, adaptation policy should be a high priority. Also, while in this chapter many methods were discussed for identifying adaptation policies, they all rely on highly uncertain impact estimates which makes policy identification difficult. As it is also considered that the countries of SEA share many specific climate change vulnerabilities the argument is made that pooling resources throughout the region can enhance the chances of identifying adaptation policies. The recommendation is made that a regional institution be used to combine the scientific and economic resources for the task of identifying adaptation policies. For reasons such as its resources, capabilities and proven record on environmental issues it is concluded that ASEAN be promoted as the institution with the responsibility for the coordination of adaptation policy identification in SEA.

8
Conclusion: Major Contributions and Recommendations for Future Research

8.1 Introduction

The main objective of this book is to analyse the economic implications of climate change for SEA, in particular as they relate to climate change adaptation. To illustrate, a summary of the means by which the objective was achieved is provided. It began by identifying the geographical scope of the book and the climate change and economic characteristics of SEA. This led to the estimation of the impacts of climate change for the region. These impact estimates were used in the implementation of the SEADICE model. This model sought to represent the dynamic optimal outcomes of the interaction between economic growth and climate change for the region. After an analysis of the theories behind adaptation to climate change the conclusions reached were used to incorporate autonomous adaptation into the SEADICE model. Given the results of the model and the arguments made in the book, it ends with relevant discussions and policy recommendations for both mitigation and adaptation for SEA. Rather than repeat these arguments in detail, this chapter concludes the book by highlighting the contributions and the limitations of this book, then it finishes with some consideration of the future areas of research this book may encourage.

8.2 Major contributions to the literature

This book has covered several distinct areas of climate change economics including macroeconomic impact estimates, climate change economic modelling and the theoretical and practical aspects of the economics of adaptation to climate change. All of these areas of the economics of climate change are distinguished by the fact that they are all subject to great uncertainties. This is a fact that must always be considered and properly acknowledged in books of this type. The major findings and contributions arising from the book will be the focus of this section.

The main goal of Chapter 3 was to estimate the aggregate economic impact of 2×CO_2 climate change for SEA. This particular area of climate change economics has focused to a large extent on estimates for the United States. In fact to the author's knowledge no estimate of this type has ever been undertaken for SEA.[38] Therefore, the impact estimate represents a contribution to the literature. A contribution which is important for these reasons:

1. Developing regions are thought to be the most vulnerable to climate change. Therefore, climate change impacts are likely to constitute a larger percentage of GDP and are therefore of more concern.
2. Estimates of this type are one of the steps required to estimate the overall vulnerability of the region (Amadore et al. 1996; Kelly and Adger 1997).
3. This aggregate climate change impact estimate can provide data for regional or global climate change models and for research into climate change adaptation where estimates of the potential future benefits (damage prevented) of adaptation are needed.

The main contribution from Chapter 4 is similar to that of Chapter 3 because it primarily results from the uniqueness of the book's geographical scope. A dynamic optimisation model using parameter values specific to SEA was implemented in Chapter 4 to examine the dynamic relationships between the economy and climate change for the region. Very few models of this type have been created outside of developed regions. Consequently the SEADICE model implemented here is a contribution to the literature. It supports the work of Islam (1994) by using the DICE modelling framework to represent a region of the globe, and its successful implementation further demonstrates the robustness of this method. Again, this type of contribution is significant because the developing countries have been substantially underrepresented, yet they are likely to be the regions most vulnerable to the effects of climate change.

The main goal of Chapter 5 was to provide a comprehensive coverage of the theories behind climate change adaptation and to develop a framework that could be used in the confines of economic theory. The major contribution of this chapter is the incorporation of endogenous growth into a dynamic optimisation model in order to represent autonomous adaptation to climate change. Throughout the literature specifications of autonomous adaptation to climate change have not been attempted. In order to represent autonomous adaptation it is assumed that the level of technology is a determinant of the level of autonomous adaptation. A solution incorporating this assumption was then found by using a technique of new growth theory known as endogenous technical progress. A literature review revealed that only two aggregate economic-climate models have attempted to incorporate climate change adaptation, and neither of those were optimisation models. The application of endogenous growth theory to an optimal growth model followed the method employed by Islam (1996), where endogenous technical progress is represented using the Shell (1966) equation. In the Islam (1996) experiment, however, the context did not relate to autonomous adaptation. The unique contribution of this book results from the assumption that this method can provide a representation of autonomous adaptation to climate change. This type of application has never been attempted before. Chapter 5 provides a framework where the representation of autonomous adaptation is now possible in economic models of climate change.

While not as significant as the modelling section of this book the final two chapters provide policy prescriptions based at least partly on the earlier contributions. In Chapter 6 it is recommended that SEA pursue opportunities resulting from the imminent demand for CDM partners from Annex I countries. SEA's unique position with respect to its emission profile and stage of economic development allows it to take advantage of a comparative advantage over other regions. In Chapter 7 it is recommended that a regional approach be taken to adaptation policy, namely the identification of adaptation options. ASEAN is recommended as the institution in the region that is the most capable of coordinating this type of activity. While the policy recommendations for both mitigation and adaptation are not controversial, they do represent a contribution because in this instance they have been supported by research based on the unique modelling results contributed by this book.

8.3 Limitations of the book

The most significant limitation of this book, shared with any study involving climate change, is the high level of uncertainty behind the science and economics of climate change. This field of research is inherently uncertain due to the long time periods involved and the large gaps in knowledge for the newly developing science and economics research areas that the problem of climate change has created. It could conceivably be decades before the true nature of climate change is known with enough certainty. Kelly et al. (2000) use a simple reduced form climate model and historical temperature records to estimate that it could take over 50 years until there is sufficient confidence in the knowledge of temperature change due to climate change. However, the problem is so important and so potentially dangerous that action is required despite the uncertainties involved by following the precautionary principle. The results presented in this book are highly uncertain, however, the mistake should not be made to interpret climate scenarios as predictions, instead they should be seen as discrete descriptions characterised by plausibility, as opposed to probability (Page 2001). The same argument applies to this book; the most plausible results are presented, given current state of the art knowledge.

The other major limitation of the book is the use of the SEADICE model. The SEADICE model bears the same criticisms as all aggregated IAM models such as the DICE model. These criticisms were acknowledged in Chapter 4, and more particularly in Sections 4.2 and 4.3.3. While these criticisms remain, IAMs continue to be used for the analysis of various climate change issues. This is because, despite their limitations, IAMs serve a purpose. As Toth notes, 'If the building blocks are so shabby, is it worthwhile building integrated models at all? The answer is clearly yes, despite the present weaknesses of the models. The reason is that modeling forces us to reveal our assumptions and changing those assumptions shows how important they are with respect to the outcome.' (Toth 1995, p. 265)

8.4 Areas for further research

Research cannot only answer questions, but lead to new questions that can be answered by further research. The number of possible areas for future research for the issue of climate change are significant as a result of the research area being so new and advancing rapidly over time. The contributions made in this book also highlight possible future directions in climate change research. Two in particular stand out.

An obvious area for further research is in the estimation of aggregate climate change damages. As state of the art research is completed on the various sectoral impacts of climate change the opportunity is created to also improve aggregate impact assessments. It is hoped that the model implemented in this book may act as a base for further refinements and improvements so that the understanding of the economic impact of climate change on SEA is enhanced. For instance if a more advanced estimate of the agricultural impact of climate change on SEA is made then it can be incorporated with the other sectoral data compiled in this book to arrive at an improved aggregate impact estimate.

Chapter 5 provides a framework where the representation of autonomous adaptation is now possible in economic models of climate change. It has been assumed in this book that technology is a determinant of autonomous adaptation. The assumption at this stage is based on the consensus view held by the IPCC (1996b) that technology is one of the determinants of autonomous adaptation. However, the significance of this relationship is unknown at this stage. Determining the extent of this relationship and grounding it in a more sophisticated economic framework would be another interesting subject for further research.

8.5 Conclusion

Any area of research that is developing rapidly is very challenging to undertake, but also very rewarding. The analysis of the economic impacts of climate change and the possible adaptations to it at the macroeconomic level is a scholarly pursuit that is still evolving. It is easy to write off the problem of global climate change as too complex, too uncertain or too large. However, the potential impact is of such magnitude that it simply cannot be ignored. Many serious questions still have not been answered and many more have not even been asked as yet about the economic effects of climate change. This book is an attempt to further explore aspects of climate change economics that have been neglected and are only recently receiving recognition as legitimate for research, in particular the aggregate economic impacts of climate change on a developing region and the economic representation of adaptation to climate change.

Notes

1 When climate change is mentioned throughout the book it refers to human-induced or anthropogenic climate change. Climate change that occurs naturally (e.g. the Ice Age) will be described as natural climate change.

2 Principle 15 of the Rio Declaration from the 1992 United Nations Conference on Environment and Development, also known as Agenda 21 states:

> 'In order to protect the environment, the precautionary approach shall be widely applied by States according to their capabilities. Where there are threats of serious or irreversible damage, lack of full scientific certainty shall not be used as a reason for postponing cost-effective measures to prevent environmental degradation.' (United Nations 1992, Article 15).

3 Although, it must be noted that at the time he did dismiss this factor as insignificant and temporary and instead postulated that volcanic eruptions were the major contributing factor to CO_2 emissions and climate change (Arrhenius 1896).

4 Other early literature that referred to the potential for climate change includes Callendar (1938, 1949, 1958, 1961), Chamberlin (1897, 1898, 1899) and Plass (1956a, 1956b, 1956c, 1961).

5 The Annex I countries consist of the OECD group and the economies in transition of Central and Eastern Europe and the former Soviet Union.

6 No regrets measures are defined as policy actions that have climate change benefits but otherwise should be implemented as they have no net costs.

7 The actual average emission reduction is closer to 10% because many of the Annex I countries have not succeeded in meeting the earlier non-binding agreement of 1990 emission levels by the year 2000.

8 South East Asia in this dissertation includes the following countries: Cambodia, Indonesia, Laos, Malaysia, Myanmar, Philippines, Singapore, Thailand and Vietnam.

9 Doi Moi, was coined in 1986 by the Vietnamese Communist Party for their reform of the economy by way of a transition from a centrally planned Stalinist command economy to a market economy with socialist direction, commonly referred to as market socialism.

10 These national communications are all available for download from the UNFCCC website at www.unfccc.int.

11 In Indonesia the forests remove over twice as much CO_2 as they emit, resulting in negative overall emissions of CO_2 (ALGAS 1998a).

12 The original Kuznets (1955, 1963) Curve was proposed as a hypothesis that initially the distribution of income may become more unequal but later improve through the development process as incomes rise. Further research can be found in Roberts and Grimes (1997); Agras and Chapman (1999); List and Gallet (1999) and Magnani (2000).
13 There is an increase in the indifference for the environmental consequences of economic development.
14 Figures 2.4–2.7 were compiled using emissions data from Marland et al. (1999) and economic data from Heston and Summers (1995) for the period 1950–91.
15 Most studies use the term damage when referring to the outcome of the effects of climate change. This however can be misleading as climate change can have both negative and positive effects on the economy, e.g. higher temperature in sub-Arctic regions have the potential to substantially increase crop yields. Therefore, to provide more accurate terminology the term *impact* will be used throughout the dissertation instead of *damage*.
16 Sites used were in Bangladesh, India, Indonesia, Malaysia, Myanmar, Philippines, Thailand, China, Japan, South Korea and Taiwan from 68 sites. Scenarios were based on the GFDL, GISS and UKMO models, using two models of rice growth ORYZA1 and SIMRIW.
17 Economic impacts on SEA nations of natural disasters were over ten times larger than for developed nations such as Canada and Australia (IFRCRCS 1997).
18 For an in-depth discussion of this issue see Albala-Bertrand (1993).
19 In any case the climate module of the DICE model is calibrated against a more complex climate model and follows the results of the more complex model very closely (Nordhaus and Boyer 2000).
20 For more information on the GAMS program see Brooke, Kendrick and Meeraus (1992).
21 Based on Sanderson and Islam (2001).
22 After the $2\times CO_2$ level, the quadratic nature of the damage function means that the differences will become much greater over time. However, estimates past $2\times CO_2$ are rarely made, as they are even more speculative than $2\times CO_2$ estimates.
23 Which is the way humans and other animals employ learning rules to adapt their behavior to environmental conditions.
24 'Adaptation is concerned with responses to both the adverse and positive effects of climate change. It refers to any adjustment whether passive, reactive, or anticipatory that can respond to anticipated or actual consequences associated with climate change.' (IPCC 1996b, p. 831).
25 Other definitions of climate change adaptation include:

> 'Adaptation to climate is the process through which people reduce the adverse effects of climate on their health and well-being, and take advantage of the opportunities that their climatic environment provides.' (Burton 1992 quoted in Feenstra et al. 1998).

'... the term adaptation means any adjustment, whether passive, reactive or anticipatory, that is proposed as a means for ameliorating the anticipated adverse consequences associated with climate change.' (Stakhiv 1993 quoted in Smit et al. 2000).

'Adaptability refers to the degree to which adjustments are possible in practices, processes, or structures to projected or actual changes of climate. Adaptation can be spontaneous or planned, and can be carried out in response to or in anticipation of changes in conditions.' (IPCC 1996b, p. 831).

26 Others such as Carter (1996) have identified several strands of in-built adjustments within the definition of autonomous adaptation: unconscious or automatic reactions to climate change, routine adjustments which are everyday conscious responses to climate change and tactical adjustments which are higher level responses which require a behavioral change. These types of definitions, while intriguing, are too detailed to be of use for the type of economic modelling attempted in this dissertation.
27 Equations based on Fankhauser, Smith and Tol (1999).
28 Using an analogy, it is reasonable to expect that people will not board up their houses and leave town if they are informed that this summer there will be a 5% chance of a cyclone. People weigh their costs and benefits of actions and in the majority of cases will not act until the potential costs outweigh benefits. In the cyclone case it is most likely that evacuation is not undertaken until at least some physical evidence of the cyclone is available.
29 Change can occur in the information the agent has available and the physical effects of climate change.
30 The Inada conditions need the marginal product of the factors of production to approach zero when their use goes to infinity and vice versa.
31 According to Agenda item 3b (iii) Section C Paragraph 5 of the COP7 documentation (available online at: www.unfccc.int) the Executive Board of the CDM is responsible for among other things:

- Supervision of the CDM, under the authority and guidance of the COP.
- Making recommendations on further modalities and procedures for the CDM, as appropriate.
- Approving new methodologies related to, *inter ali*a, baselines, monitoring plans and project boundaries.
- Be responsible for the accreditation of operational entities.
- Develop, maintain and make publicly available a repository of approved rules, procedures, methodologies, standards and maintain a publicly available database of CDM project activities.

32 The author is aware that the imposition of a carbon tax across the SEA region is very unlikely as the policy is more suited to national policy action. The results presented are only meant to be illustrative, just as

those given for global carbon tax estimates in past studies (Nordhaus 1994a).

33 The Philippines and Malaysia were the first to ratify the UNFCCC in October 1994, followed by Indonesia in November 1994, Myanmar and Vietnam in February 1995, Thailand in March 1995, Laos in April 1995, Cambodia in March 1996 and Singapore in August 1997.

34 An in-depth discussion of the issues of positive and normative analysis as they apply to climate change adaptation can be found in Smit et al. (1999).

35 Individual researchers have also offered their own criteria and procedures for adaptation policy assessments, see Smith (1997), and Smith, Ragland and Pitts (1996) for more detail.

36 For more information on sustainable economic development see Bossel (1999); Faucheux, Pearce and Proops (1996); Munasinghe (1993, 2001).

37 The author is aware that other institutions in the region such as ADB and APEC could also fulfill a role similar to that proposed here for ASEAN.

38 The closest example can be found in Tol (1996), where an estimate was made for South and Southeast Asia. This regional grouping did not include Vietnam and Laos and included countries from South Asia.

Bibliography

ADB (Asian Development Bank) (1991), *Asian Development Outlook 1991*, Asian Development Bank, Manila.

ADB (Asian Development Bank) (1994), *Climate Change in Asia: Indonesia Country Report on Socioeconomic Impacts of Climate Change and a National Response Strategy*, Asian Development Bank, Manila.

ADB (Asian Development Bank) (1999), *Annual Report 1999*, Tien Wah Press, Singapore.

ADB (Asian Development Bank) (2000), *Environments in Transition: Cambodia Lao PDR Thailand Viet Nam*, Asian Development Bank, Manila.

ALGAS (Asia Least-Cost Greenhouse Gas Abatement Strategy) (1998a), *Indonesia: Executive Summary*, Asian Development Bank Global Environment Facility United Nations Development Programme, Manila.

ALGAS (Asia Least-Cost Greenhouse Gas Abatement Strategy) (1998b), *Philippines: Executive Summary*, Asian Development Bank Global Environment Facility United Nations Development Programme, Manila.

ALGAS (Asia Least-Cost Greenhouse Gas Abatement Strategy) (1998c), *Thailand: Executive Summary*, Asian Development Bank Global Environment Facility United Nations Development Programme, Manila.

ALGAS (Asia Least-Cost Greenhouse Gas Abatement Strategy) (1998d), *Viet Nam: Executive Summary*, Asian Development Bank Global Environment Facility United Nations Development Programme, Manila.

Agras, J. and Chapman, D. (1999), 'A Dynamic Approach to the Environmental Kuznets Curve Hypothesis', *Ecological Economics*, vol. 28, pp. 267–77.

Albala-Bertrand, J.M. (1993), *Political Economy of Large Natural Disasters: With Special Reference to Developing Countries*, Clarendon Press, Oxford.

Amadore, L. Bolhofer, W.C. Cruz, R.V. Feir, R.B. Freysinger, C.A. Guill, S. Jalal, K.F. Iglesias, A. Jose, A. Leatherman, S. Lenhart, S. Mukherjee, S. Smith, J.B. and Wisniewski, J. (1996), 'Climate Change Vulnerability and Adaptation in Asia and the Pacific: Workshop Summary', *Water Air and Soil Pollution*, vol. 92, no. 1/2, pp. 1–12.

Amien, I. Rejekiningrum, P. Pramudia, A. and Susanti, E. (1996), 'Effects of Interannual Climate Variability and Climate Change on Rice Yield in Java Indonesia', *Water Air and Soil Pollution*, vol. 92, no. 1/2, pp. 29–39.

Anand, S. and Sen, A. (1996), *Sustainable Human Development: Concepts and Priorities*, UNDP Office of Development Studies, No. 1, Discussion Paper Series, New York.

Anderson, J.W. (1997), *Climate Change Clinton and Kyoto: The Negotiations over Global Warming*, Working Paper, Resources for the Future.

Angel, D.P. and Rock, M.T. (2000), *Asia's Clean Revolution: Industry Growth and the Environment*, Greenleaf, Sheffield.

Arrhenius, S. (1896), 'The Influence of the Carbonic Acid in the Air Upon the Temperature of the Ground', *Philosophical Magazine*, vol. 41.

Arrow, K.J. (1962), 'The Economic Implications of Learning-by-Doing', *Review of Economic Studies*, vol. 29, pp. 155–73.

Arrow, K.J. Cline, W.R. Mäler, K.G. Munasinghe, M. Squitieri, R. and Stiglitz J.E. (1996), 'Intertemporal Equity, Discounting, and Economic Efficiency', in Bruce, J.P. Lee, H. and Haites, E.F. (eds.), *Climate Change 1995: Economic and Social Dimensions of Climate Change*, Cambridge University Press, Cambridge.

Asia Pacific Energy Research Centre (2001), *Making the Clean Development Mechanism Work: With Some Case Studies in the APEC Region*, Tokyo, Japan.

Ausubel, J.H. (1993), 'Mitigation and Adaptation for Climate Change: Answers and Questions', *The Bridge*, vol. 23, no. 3, pp. 15–30.

Ausubel, J.H. (1995), 'Technical Progress and Climatic Change', *Energy Policy*, vol. 23, no. 4/5, pp. 411–16.

Azar, C. (1995), *Long-Term Environmental Problems Economic Measures and Physical Indicators,* Chalmers University of Technology, Sweden.

Azar, C. and Dowlatabadi, H. (1999*), A Review of Technical Change in Assessment of Climate Policy*, Centre for Integrated Study of the Human Dimensions of Global Change.

Baker, P.T. (1984), 'The Adaptive Limits of Human Populations', *Man*, vol. 19, pp. 1–14.

Barro, R.J. and Sala-i-Martin, X. (1995), *Economic Growth,* McGraw-Hill, New York.

Bartelmus, P. (1999), 'Green Accounting for a Sustainable Economy: Policy Use and Analysis of Environmental Accounts in the Philippines', *Ecological Economics*, vol. 29, pp. 155–70.

Bautista, G.M. (1990), 'The Forestry Crisis in the Philippines: Nature, Causes and Issues', *The Developing Economies*, vol. 28, no. 1, pp. 67–94.

Bazzaz, F. and Sombroek, W. (eds), (1996), *Global Climate Change and Agricultural Production: Direct and Indirect Effects of Changing Hydrological Pedological and Plant Physiological Processes,* John Wiley & Sons, Chichester.

Benhaim, J. and Schembri, P. (1996), 'Technical Change: An Essential Variable in the Choice of a Sustainable Development Trajectory', in Faucheux, S. Pearce, D. and Proops, J. (eds) *Models of Sustainable Development,* Edward Elgar, Cheltenham.

Benioff, R. Guill, S. and Lee, J. (eds), (1996), *Vulnerability and Adaptation Assessments: An International Handbook,* Kluwer Academic Publishers, Dordrecht.

Bhattacharya, S.C. Pittock, A.B. and Lucas, N.J.D. (eds) (1994), *Global Warming Issues in Asia,* Proceedings of the Workshop on Global Warming Issues in Asia 8–10 September 1993, Regional Energy Resources Information Center, Bangkok.

Bijlsma, L. Ehler, C.N. Klein, R.J.T. Kulshrestha, S.M. McLean, R.F. Mimura, N. Nicholls, R.J. Nurse, L.A. Perez Nieto, H. Stakhiv, E.Z. Turner, R.K. Warrick, R.A. (1996), 'Coastal Zones and Small Islands', in Watson, R.T. Zinyowera, M.C. and Moss, R.H. *Impacts Adaptations and Mitigation of*

Climate Change: Scientific-Technical Analyses, Cambridge University Press, Cambridge.

Boer, B. Ramsey, R. and Rothwell, D.R. (1998), *International Environmental Law in the Asia Pacific*, Kluwer Law International, London.

Boonpragob, K. and Santisirisomboon, J. (1996), 'Modeling Potential Changes in Forest Area in Thailand Under Climate Change', *Water, Air and Soil Pollution*, vol. 92, pp. 107–17.

Bossel, H. (1999), *Indicators for Sustainable Development: Theory, Method, Applications*, International Institute for Sustainable Development, Canada.

Brack, D. and Grubb, M. (1996), 'Climate Change: A Summary of the Second Assessment Report of the IPCC', *Briefing Paper no. 32*, The Royal Institute of International Affairs, London.

Brandon, C. and Ramankutty, R. (1993), *Toward an Environmental Strategy for Asia*, Discussion Paper no. 224, The World Bank, Washington.

Brogan, P. Kennedy, J.J. Harris, I. et al. (2006), 'Uncertainty Estimates in Regional and Global Observed Temperature Changes: A New Dataset from 1850', *Journal of Geophysical Research*, vol. 111.

Brooke, A. Kendrick, D. and Meeraus, A. (1992), *GAMS A User's Guide*, Boyd & Fraser, Massachusetts.

Brookfield, H. and Byron, Y. (eds) (1993), *South-East Asia's Environmental Future: The Search for Sustainability*, United Nations University Press, Tokyo.

Broome, J. (1992), *Counting the Costs of Global Warming*, The White Horse Press, Cambridge.

Bruce, J.P. (1999), 'Disaster Loss Mitigation as an Adaptation to Climate Variability and Change', *Mitigation and Adaptation Strategies for Global Change*, vol. 4, no. 3/4, pp. 295–306.

Budyko, M.I. (1996), 'Past Changes in Climate and Societal Adaptations', in Smith, J.B. Bhatti, N. Menzhulin, G.V. Benioff, R. Campos, M. Jallow, B. Rijsberman, F. Budyko, M.I. and Dixon, R.K. (eds) *Adapting to Climate Change: An International Perspective*, Springer-Verlag, New York, pp. 16–26.

Burby, R.J. and Nelson, A.C. (1991), 'Local Government and Public Adaptation to Sea-Level Rise', *Journal of Urban Planning and Development*, vol. 117, no. 4, pp. 140–53.

Burton, I. (1992), *Adapt and Thrive*, Canadian Climate Centre, Downsview, Ontario, Unpublished manuscript.

Burton, I. (1996), 'The Growth of Adaptation Capacity: Practice and Policy', in Smith, J.B. Bhatti, N. Menzhulin, G.V. Benioff, R. Campos, M. Jallow, B. Rijsberman, F. Budyko, M.I. and Dixon, R.K. (eds), *Adapting to Climate Change: An International Perspective*, Springer-Verlag, New York, pp. 55–67.

Burton, I. (1997), 'Vulnerability and Adaptive Response in the Context of Climate and Climate Change', *Climatic Change*, vol. 36, pp. 185–96.

Callendar, G.S. (1938), 'The Artificial Production of Carbon Dioxide and its Influence on Temperature', *Quarterly Journal of the Royal Meteorological Society*, vol. 64, pp. 223–37.

Callendar, G.S. (1949), 'Can Carbon Dioxide Influence Climate?', *Weather*, vol. 4, pp. 310–14.

Callendar, G.S. (1958), 'On the Amount of Carbon Dioxide in the Atmosphere', *Tellus,* vol. 10, pp. 243–8.

Callendar, G.S. (1961), 'Temperature Fluctuations and Trends over the Earth', *Quarterly Journal of the Royal Meteorological Society,* vol. 87, pp. 1–12.

Cameron, O. (1996), 'Japan and South-East Asia's Environment', in Parnwell, M.J.G. and Bryant, R.L. (eds), *Environmental Change in South-East Asia,* Routledge, New York, pp. 67–94.

Carter, T.R. (1996), 'Assessing Climate Change Adaptations: The IPCC Guidelines', in Smith, J.B. Bhatti, N. Menzhulin, G.V. Benioff, R. Campos, M. Jallow, B. Rijsberman, F. Budyko, M.I. and Dixon, R.K. (eds), *Adapting to Climate Change: An International Perspective,* Springer-Verlag, New York, pp. 27–43.

Carter, T.R. Parry, M.L. Harasawa, H. and Nishioka, N. (1994), *IPCC Technical Guidelines for Assessing Climate Change Impacts and Adaptations,* University College London, London.

Centre for Strategic Economic Studies (1998), *Crisis in East Asia: Global Watershed or Passing Storm?* Conference Report, March 10–11, CSES, Victoria University of Technology, Melbourne.

Chamberlin, T.C. (1897), 'A Group of Hypotheses Bearing on Climatic Changes', *Journal of Geology,* vol. 5, pp. 653–83.

Chamberlin, T.C. (1898), 'The Influence of Great Epochs of Limestone Formation upon the Constitution of the Atmosphere', *Journal of Geology,* vol. 6, pp. 609–21.

Chamberlin, T.C. (1899), 'An Attempt to Frame a Working Hypothesis of the Cause of Glacial Periods on an Atmospheric Basis', *Journal of Geology,* vol. 7, pp. 545–84, 667–85, 751–87.

Chan, N.W. (1995), 'Flood Disaster Management in Malaysia: An Evaluation of the Effectiveness of Government Resettlement Schemes', *Disaster Prevention and Management,* vol. 4, no. 4, pp. 22–9.

Chapman, D. Suri, V. and Hall, S.G. (1995), 'Rolling DICE for the Future of the Planet', *Contemporary Economic Policy,* vol. 8, pp. 1–9.

Choe, K.A. Whittington, D. and Lauria, D.T. (1996), 'The Economic Benefit of Surface Water Quality Improvements in Developing Countries: A Case Study of Daveo Philippines', *Land Economics,* vol. 72, no. 4, pp. 519–37.

Chou, L.M. (ed.) (1994), *Implications of Expected Climate Changes in the East Asian Seas Region: an Overview,* RCU/EAS Technical Report series No. 2, United Nations Environmental Programme, Bangkok.

CIESIN (Consortium for International Earth Science Information Network) (1995), *Thematic Guide to Integrated Assessment Modeling of Climate Change,* University Center, Michigan, available online at: http://sedac.ciesin.org/mva/iamcc.tg/TGHP.html

Cline, W. (1992), *The Economics of Global Warming,* Institute of International Economics, Washington.

Cortright, J. (2001), *New Growth Theory Technology and Learning: A Practitioners Guide,* Reviews of Economic Development Literature and Practice, no. 4, Impresa Inc., Portland.

Costanza, R. (1996), *Managing the Global Commons: Review*, available online at: http://web.umr.edu/~rrbryant/econ340/ManagingthecommonsReview.pdf.

Cypher, J.M. and Dietz, J.L. (1997), *The Process of Economic Development*, Routledge, London.

Darmastadter, J. and Toman, M.A. (eds) (1993), *Assessing Surprises and Nonlinearities in Greenhouse Warming: Proceedings of an Interdisciplinary Workshop*, Resources for the Future, Washington.

Darwin, C. (1859), *On the Origin of Species*, Harvard University Press, London.

Darwin, R. Tsigas, M. Lewandrowski, J. and Ranese, A. (1995), *World Agriculture and Climate Change: Economic Adaptations*, United States Department of Agriculture, Economic Research Service, AER-703, Washington DC.

Darwin, R. Tsigas, M. Lewandrowski, J. and Ranese, A. (1996), 'Land Use and Cover in Ecological Economics', *Ecological Economics*, vol. 17, pp. 157–81.

De Canto, S.J. Howarth, R.B. Sanstad, A.H. and Thompson, S.L. (2000), *New Directions in the Economics and Integrated Assessment of Global Climate Change*, Pew Center on Global Climate Change, Battelle.

Decision Focus (1996), *Tool for Environmental Assessment and Management: Quick Reference Pamphlet*. Prepared by Decision Focus Incorporated in co-operation with Narayanan V. of Technical Resources International and Office of Policy, Planning and Evaluation, United States Environmental Protection Agency, Washington, DC.

DEFRA (2005), *Climate Change and the Greenhouse Effect: A Briefing from the Hadley Centre*, available at: http://www.metoffice.com/research/hadleycentre/pubs/brochures/2005/climate_greenhouse.pdf.

Delft Hydraulics (1993), *Sea Level Rise: A Global Vulnerability Assessment*, 2nd Revised Edition, Delft Hydraulics and Ministry of Transport, Public Works and Water Management, The Hague.

Dixon, C. (1991), *South East Asia in the World Economy*, Cambridge University Press, Cambridge.

Dixon, R.K. (1999), 'Special Issue of IPCC Workshop on Adaptation to Climate Variability and Change: Methodological Issues', *Mitigation and Adaptation Strategies for Global Change*, vol 4. nos. 3–4.

Dopfer, K. (1979), *The New Political Economy of Development: Integrated Theory and Asian Experience*, Macmillan Press, London.

Dowlatabadi, H. and Morgan, M.G. (1993), 'A Model Framework for Integrated Studies of the Climate Problem', *Energy Policy*, vol. 21, pp. 209–21.

Drysdale, P. and Huang, Y. (1995), 'Growth, Energy and the Environment: New Challenges for the Asian-Pacific Economy', *Asian-Pacific Economic Literature*, vol. 9, no. 2, pp. 1–12.

Dubos, R. (1965), *Man Adapting*, Yale University Press, New Haven.

Edmonds, J. Roop, J.M. and Scott, M.L. (2000), *Technology and the Economics of Climate Change Policy*, Pew Center on Global Climate Change, Battelle.

England, R.W. (ed.), (1994), *Evolutionary Concepts in Contemporary Economics*, The University of Michigan Press, Michigan.

Erda, L. Bolhofer, W. Huq, S. Lenhart, S.K. Smith, J.B. and Wisniewski, J. (eds) (1996), *Climate Change Variability and Adaptation in Asia and the Pacific*, Kluwer Academic Publishers, Dordrecht.

Escano, C.R. and Buendia, L.V. (1994), 'Climate Impact Assessment for Agriculture in the Philippines: Simulation of Rice Yield Under Climate Change Scenarios', in Rosenzweig, C. and Iglesias, A. (eds), *Implications of Climate Change for International Agriculture: Crop Modeling Study*, Unites States Environmental Protection Agency, Philippines chapter, Washington, DC, pp. 1–13.

Fankhauser, S. (1993), 'The Economic Costs of Global Warming: Some Monetary Estimates', in Kaya, Y. Nakicenovic, N. Nordhaus, W.D. and Toth, F.L. (eds), *Costs Impacts and Benefits of CO_2 Mitigation*, International Institute for Systems Analysis, Laxenburg.

Fankhauser, S. (1994a), 'The Social Costs of Greenhouse Gas Emissions: An Expected Value Approach', *The Energy Journal*, vol. 15, no. 2, pp. 157–84.

Fankhauser, S. (1994b), 'The Economic Costs of Global Warming Damage: A Survey', *Global Environmental Change*, vol. 4, no. 4, pp. 310–19.

Fankhauser, S. (1995a), 'Protection Versus Retreat: The Economic Costs of Sea-Level Rise', *Environment and Planning*, vol. 27, pp. 299–319.

Fankhauser, S. (1995b), *Valuing Climate Change: The Economics of the Greenhouse*, Earthscan, London.

Fankhauser, S. (1996), 'The Potential Costs of Climate Change Adaptation', in Smith, J.B. Bhatti, N. Menzhulin, G.V. Benioff, R. Campos, M. Jallow, B. Rijsberman, F. Budyko, M.I. and Dixon, R.K. (eds), *Adapting to Climate Change: An International Perspective*, Springer-Verlag, New York, pp. 80–96.

Fankhauser, S. (1998), *The Costs of Adapting to Climate Change*, Working Paper no. 16, Global Environment Facility.

Fankhauser, S. and Tol, R.S.J. (1995), *The Social Costs of Climate Change: The IPCC Second Assessment Report and Beyond*, Institute for Environmental Studies, Amsterdam.

Fankhauser, S. and Tol, R.S.J. (1996), 'Climate Change Costs: Recent Advancements in the Economic Assessment', *Energy Policy*, vol. 24, no. 7, pp. 665–73.

Fankhauser, S. Smith, J. and Tol, R.S.J. (1999), 'Weathering Climate Change: Some Simple Rules to Guide Adaptation Decisions', *Ecological Economics*, vol. 30, pp. 427–49.

Fankhauser, S. Tol, R.S.J. and Pearce, D.W. (1997), 'The Aggregation of Climate Change Damages: A Welfare Theoretic Approach', *Environmental and Resource Economics*, vol. 10, pp. 249–66.

Faucheux, S. Pearce, D. and Proops, J. (eds) (1996), *Models of Sustainable Development*, Edward Elgar, Cheltenham.

Fedderke, J. (2001), *Technology Human Capital Growth and Institutional Development: Lessons from Endogenous Growth Theory?* Policy Paper no. 13, Econometric Research Southern Africa, University of Witwatersrand.

Feenstra, J.A. Burton, I. Smith, J.B. and Tol, R.S.J. (eds) (1998), *Handbook on Methods for Climate Change Impact Assessments and Adaptation Strategies*, version 2.0, United Nations Environmental Programme.

Felipe, J. (1997), *Total Factor Productivity Growth in East Asia: A Critical Survey*, EDRC Report Series No 65, Economics and Development Resource Centre, Asian Development Bank, Manila.

Fischer, G. Frohberg, K. Parry, M.L. and Rosenzweig, C. (1996), 'The Potential Effects of Climate Change on World Food Production and Security', pp. 199–235, in Bazzaz, F. and Sombroek, W. (eds), *Global Climate Change and Agricultural Production. Direct and Indirect Effects of Changing Hydrological, Pedological and Plant Physiological Processes,* FAO and John Wiley & Sons.

Food and Agriculture Organization of the United Nations (1996), *The State of Food and Agriculture, Food Security: Some Macroeconomic Dimensions,* Food and Agriculture Organization, Rome.

Forsyth, T. (1999), *International Investment and Climate Change: Energy Technologies for Developing Countries,* Royal Institute of International Affairs, Earthscan, London.

Fourier, J. (1824), 'Remarques Générales sur la Température du Globe Terrêstre et des Espaces Planétaires', *Annals de Chimie et de Physique,* vol. 27, pp. 136–67.

Frisancho, R. (1993), *Human Adaptation and Accommodation,* University of Michigan Press, Ann Arbor.

Frisvold, G. and Kuhn, B. (1999), *Global Environmental Change and Agriculture: Assessing the Impacts,* Edward Elgar, Cheltenham.

Ghosh, P. (ed.) (2000), *Implementation of the Kyoto Protocol: Opportunities and Pitfalls for Developing Countries,* Asian Development Bank, Manila.

Global Environment Facility (1996), *Operational Strategy,* Global Environment Facility, Washington.

Global Environment Facility (2000), *Review of Climate Change Enabling Activity Projects,* GEF Report, Washington.

Go, F.M. and Jenkins, C.L. (eds) (1997), *Tourism and Economic Development in Asia and Australasia,* Cassell Imprint, London.

Grove, R.H. Damodaran, V. and Sangwan, S. (1998), *Nature and the Orient: the Environmental History of South and Southeast Asia,* Oxford University Press, Oxford.

Grubb, M. Chapuis, T. and Duong, M.H. (1995), 'The Economics of Changing Course: Implications of Adaptability and Inertia for Optimal Climate Policy', *Energy Policy,* vol. 23, no. 4/5, pp. 417–32.

Guha, A.S. (1981), *An Evolutionary View of Economic Growth,* Clarendon Press, Oxford.

Haavelmo, T. (1954), *A Study of the Theory of Economic Evolution,* North Holland, Amsterdam.

Habito, C.F. (1993), 'A Look at the Agri-industrial Development Strategy', *Philippine Development,* vol. 20, no. 3, pp. 2–6.

Hall, S. and Mabey, N. (1995), 'Econometric Modelling of International Carbon Tax Regimes', *Energy Economics,* vol. 17, no. 2, pp. 133–46.

Hamilton, D. (1991), *Evolutionary Economics: A Study of Change in Economic Thought,* Transaction Publishing, New Jersey.

Hanisch, T. (1994), *Climate Change and the Agenda for Research,* Westview Press, San Francisco.

Hanley, N. (1992), 'Are There Environmental Limits to Cost-Benefit Analysis?', *Environmental and Resource Economics,* vol. 2, pp. 33–59.

Harrod, R.F. (1939), 'An Essay in Dynamic Theory', *Economic Journal*, vol. 49, pp. 14–33.
Heston, A., and Summers, R. (1995), *Penn World Tables*, Version 5.6, The Center for International Comparisons at the University of Pennsylvania, Pennsylvania.
Higano, Y. (1984), 'Control of Environment by Quasi-Market', *Expressways and Automobiles*, vol. 27, no. 9, pp. 31–9.
Higano, Y. (1985), 'On the Exclusion Theorem', *Regional Science and Urban Economics*, vol. 15, no. 3, pp. 449–58.
Hodgson, G.M. (1997), 'Economics and Evolution and the Evolution of Economics', in Reijnders, J. (ed.), *Economics and Evolution*, Edward Elgar, Cheltenham.
Holland, J. (1995), *Hidden Order: How Adaptation Builds Complexity*, Addison-Wesley, Reading.
Holtz-Eakin, D. and Selden, T.M. (1995), 'Stoking the Fires? CO_2 Emissions and Economic Growth', *Journal of Public Economics*, vol. 57, pp. 85–101.
Hope, C. Anderson, J. and Wenman, P. (1993), 'Policy Analysis of the Greenhouse Effect: An Application of the PAGE Model', *Energy Policy*, vol. 21, pp. 327–38.
Hulme, M. Zhao, Z-C. and Jiang, T. (1994), 'Recent and Future Climate Change in East Asia', *International Journal of Climatology*, vol. 14, pp. 637–58.
Hydrometeorological Service of Vietnam (1999), *Economics of Greenhouse Gas Limitations: Country Study Series*, UNEP Collaborating Centre on Energy and Environment, Denmark.
IFRCRCS (International Federation of Red Cross and Red Crescent Societies) (1997), *World Disaster Report 1997*, Oxford University Press, Oxford.
Iglesias, A. Erda, L. and Rosenzweig, C. (1996), 'Climate Change in Asia: A Review of the Vulnerability and Adaptation of Crop Production', *Water Air and Soil Pollution*, vol. 92, no. 1/2, pp. 13–27.
IPCC (Intergovernmental Panel on Climate Change) (1994), *Preparing to Meet the Coastal Challenges of the 21st Century*, Conference Report of the World Coast Conference in Noordwijk, WHO and UNEP, Geneva.
IPCC (1996a), *Climate Change 1995: Economic and Social Dimensions of Climate Change*, Cambridge University Press, New York.
IPCC (1996b), *Climate Change 1995: Impact, Adaptations and Mitigation of Climate Change: Scientific-Technical Analysis*, Cambridge University Press, New York.
IPCC (2001a), *Climate Change 2001: The Scientific Basis*, A Report of Working Group I of the IPCC, Summary for Policymakers, available online at: http://www.ipcc.ch/pub/tar/wg1/001.htm.
IPCC (2001b), *Climate Change 2001: Impacts Adaptation and Vulnerability*, A Report of Working Group II of the IPCC, available online at: http://www.ipcc.ch/pub/tar/wg2/001.htm.
IPCC (2001c), *Climate Change 2001: Mitigation, A Report of Working Group III of the IPCC*, available online at: http://www.ipcc.ch/pub/tar/wg3/001.htm.
IPCC (2001d), *Draft Document of the Marrakesh Accord*, October 2001, available online at: http://www.unfccc.de/cop7/documents/accords_draft.pdf.

IPIECA (International Petroleum Industry Environmental Conservation Association) and UNEP (United Nations Environmental Programme) (1991), *Climate Change and Energy Efficiency in Industry*, IPIECA, London.

Islam, S.M.N. (1994), *Australian Dynamic Integrated Climate-Economy Model (ADICE)*, Centre for Strategic Economic Studies, Victoria University of Technology, Melbourne.

Islam, S.M.N. (1996), *Global Warming Endogenous Technical Progress and New Growth Theory: A Dynamic General Equilibrium Analysis of Climate Change and Economic Growth*, Research Report, Submitted to Research and Graduate Studies Committee, Victoria University.

Islam, S.M.N. (2001), *Optimal Growth Economics*, Contributions to Economic Analysis Series, North Holland, Amsterdam.

Islam, S.M.N. and Choi, S. (1998), 'Energy, Economic Growth, and the Environment: The East Asian Framework of Integrated Modelling and Policy Studies', *Pacific and Asian Journal of Energy*, vol. 9, no. 1, pp. 1–19.

Islam, S.M.N. and Jolley, A. (1996), 'Sustainable Development in Asia: the Current State and Policy Options', *Natural Resources Forum*, vol. 20, no. 4, pp. 263–79.

Islam, S.M.N. Gigas, J. and Sheehan, P. (1996), *Cost Benefit Analysis of Climate Change: Towards an Operational Decision Making Rule for Climate Change Policy*, presented at the National Conference on Climate Change, 12–13 August, Universiti Pertanian, Malaysia.

Islam, S.M.N. Sheehan, P.J. and Sanderson, J.R. (2000), 'Costs of Economic Growth: An Application of the DICE Model to Australia', *The Otemon Journal of Australian Studies*, vol. 26, pp. 119–30.

Jakeman, A.J. and Pittock, A.B. (eds) (1994), *Climate Impact Assessment Methods for Asia and the Pacific*, Proceedings of a Regional Symposium Organised by ANUTECH Pty Ltd on Behalf of the Australian International Development Assistance Bureau 10–12 March 1993, Commonwealth of Australia, Canberra.

Jalal, K.F. (1993), *Sustainable Development, Environment and Poverty Nexus*, Occasional Paper no. 7, Asian Development Bank, Manila.

Janssen, M.A. (1996), *Meeting Targets: Tools to Support Integrated Assessment Modelling of Global Change*, Cip-Genevens Koninklijke Bibliotheek, Den Haag.

Janssen, M.A. and De Vries, B. (2000), 'Climate Change Policy Targets and the Role of Technological Change', *Climatic Change*, vol. 46, pp. 1–28.

Jia, Q. (1996), *Global Warming: Contribution by and Impact on China – An Application of the DICE Model*, Dissertation, Faculty of the Graduate College, Oklahoma State University.

Kaldor, N. and Mirrles, J.A. (1962), 'A New Model of Economic Growth', *Review of Economic Studies*, vol. 29, pp. 174–90.

Kandlikar, M. and Sagar, A. (1997), *Climate Change Science and Policy: Lessons from India*, Interim Report IR-97-035, International Institute for Applied Systems Analysis, Laxenburg.

Kates, R.W. (1997), 'Climate Change 1995 – Impacts Adaptations and Mitigation', *Environment*, vol. 39, no. 9, pp. 29–33.

Kaya, Y. Nakicenovic, N. Nordhaus, W.D. and Toth, F.L. (eds) (1993), *The Costs, Impacts, and Benefits of CO_2 Mitigation,* Proceedings of a Workshop Held on 28–30 September 1992, IIASA, Laxenburg.

Kelly, D.L. and Kolstad, C.D. (1997), *Bayesian Learning Growth and Pollution*, mimeo.

Kelly, D.L. and Kolstad, C.D. (1998), 'Integrated Assessment Models for Climate Change Control', in Folmer, H. and Tietenberg, T. (eds), *International Yearbook of Environmental and Resource Economics 1999/2000: A Survey of Current Issues*, Edward Elgar, Cheltenham.

Kelly, D.L. Kolstad, C.D. Schlesinger, M.E. and Andronova, N.G. (2000), *Learning About Climate Sensitivity from the Instrumental Near-Surface Temperature Record*, Working Paper, University of California, Santa Barbara.

Kelly, P. and Adger, W.N. (1997), *Assessing Vulnerability to Climate Change and Facilitating Adaptation*, Centre for Social and Economic Research on the Global Environment Working Paper, GEC 99–07, University of East Anglia, Norwich.

Khanna, N. and Chapman, D. (1997), *A Critical Overview of the Economic Structure of Integrated Assessment Models of Climate Change*, Cornell University.

Klein, R.J.T. and Tol, R.S.J. (1997), Adaptation to Climate Change: Options and Technologies An Overview Paper, Technical Paper FCCC/TP/1997/3, United Nations Framework Convention on Climate Change Secretariat, Bonn, available online at: http://www.unfccc.int/resource/docs/tp/tp3.pdf.

Kneese, A.V. (1977), *Economics and the Environment*, Penguin Books, Middlesex.

Kovats, R.S. Menne, B., McMichael, A.J. Corvalan, C. and Bertollini, R. (2000), *Climate Change and Human Health: Impact and Adaptation*, World Health Organization, Geneva.

Krugman, P. (1996), *What Economists can Learn from Evolutionary Theorists*, available online at: http://web.mit.edu/krugman/www/evolute.html.

Kuznets, S. (1955), 'Economic Growth and Income Inequality', *American Economic Review*, vol. 49, pp. 1–28.

Kuznets, S. (1963), 'Quantitative Aspects of the Economic Growth of Nations, VII: The Distribution of Income by Size', *Economic Development and Cultural Change*, vol. 11, pp. 1–92.

Lao People Democratic Republic (2000), *The First National Communication on Climate Change*, Science Technology and Environment Agency, Vientiane.

Leary, N.A. (1999), 'A Framework for Benefit-Cost Analysis of Adaptation to Climate Change and Climate Variability', *Mitigation and Adaptation Strategies for Global Change*, vol. 4, pp. 307–18.

Leatherman, S.P. (1996), 'Shoreline Stabilization Approaches in Response to Sea Level Rise: U.S. Experience and Implications for Pacific Island and Asian Nations', *Water Air and Soil Pollution*, vol. 92, no. 1/2, pp. 149–57.

Lebel, L. and Murdiyarso, D.M. (1998), *Modelling Global Change Impacts on Tropical Landscapes and Biodiversity*, IC-SEA Report no. 5. Impacts Centre for Southeast Asia, Bogor, Indonesia.

Lebel, L. and Steffen, W.S. (eds) (1998), *Global Environmental Change and Sustainable Development: Science Plan for a SARCS Integrated Study*, Science Plan, Southeast Asian Regional Committee for START (SARCS).

Lewandrowski, J. Darwin, R.F. Tsigas, M. and Raneses, A. (1999), 'Estimating Costs of Protecting Global Ecosystem Diversity', *Ecological Economics*, vol. 29, pp. 111–25.

List, J.A. and Gallet, C.A. (1999), 'The Environmental Kuznets Curve: Does One Size Fit All?', *Ecological Economics*, vol. 31, pp. 409–23.

Lucas, R.E. (1988), 'On the Mechanics of Economic Development', *Journal of Monetary Economics*, vol. 22, no. 1, pp. 3–42.

Luo, Q. and Lin, E. (1999), 'Agricultural Vulnerability and Adaptation in Developing Countries: The Asia-Pacific Region', *Climatic Change*, vol. 43, pp. 729–43.

Mabey, N. Hall, S. Smith, C. and Gupta, S. (1997), *Argument in the Greenhouse: The International Economics of Controlling Global Warming*, Routledge, London.

Maddison, D. (1995), 'A Cost-Benefit Analysis of Slowing Climate Change', *Energy Policy*, vol. 23, no. 4/5, pp. 337–46.

Magnani, E. (2000), 'The Environmental Kuznets Curve Environmental Protection Policy and Income Distribution', *Ecological Economics*, vol. 32, pp. 431–43.

Magnussan, L. (ed.) (1994), *Evolutionary and Neo-Schumpetarian Approaches to Economics*, Kluwer, Massachusetts.

Malaysian Science and Technology Information Centre (1998), *National Survey of Research and Development*, Ministry of Science, Technology and the Environment, Kuala Lumpur.

Malik, U. (1994), 'Energy Technologies and Policies for Limiting Greenhouse Gas Emissions in Asia', in Sharma, S. (ed.), *Energy, the Environment and the Oil Market*, Institute of Southeast Asian Studies, Singapore.

Manne, A. and Richels, R. (1992), *Buying Greenhouse Insurance: The Economic Costs of Carbon Dioxide Emission Limits*, MIT Press, Cambridge.

Manne, A. Mendelsohn, R. and Richels, R. (1995), 'MERGE: A Model for Evaluating Regional and Global Effects of GHG Reduction Policies', *Energy Policy*, vol. 23, no. 1, pp. 17–34.

Marks, J.M. (1995), *Human Biodiversity: Genes, Race, and History*, Aldine de Gruyter, New York.

Marland, G. Boden, T.A. Andres, R.J. Brenkert, A.L. and Johnston, C.A. (1999), 'Global, Regional, and National Fossil Fuel CO_2 Emissions', in *Trends: A Compendium of Data on Global Change*, Carbon Dioxide Information Analysis Center, Oak Ridge National Laboratory, United States Department of Energy, Oak Ridge.

Maslin, M. (2004), *Global Warming: A Very Short Introduction*, Oxford University Press, New York.

Matsuoka, Y. Kainuma, M. and Morita, T. (1995), 'Scenario Analysis of Global Warming using the Asian Pacific Integrated Model (AIM)', *Energy Policy*, vol. 23, no. 4/5, pp. 357–71.

Matthews, R.B. Kropff, M.J. Horie, T. and Bachelet, D. (1997), 'Simulating the Impact of Climate Change on Rice Production in Asia and Evaluating Options for Adaptation', *Agricultural Systems*, vol. 54, no. 3, pp. 399–425.

Maurseth, P.B. (2001), *Recent Advancements in Growth Theory: A Comparison of Neoclassical and Evolutionary Perspectives*, Working Paper no. 615, Norwegian Institute of International Affairs.

McGregor, J.L. Katzfey, J.J. and Nguyen, K.C. (1998), *Fine Resolution Simulations of Climate Change for Southeast Asia,* Final report for a Research Project commissioned by Southeast Asian Regional Committee for START (SARCS), CSIRO Atmospheric Research, Aspendale.

McLean, R. and Mimura, N. (eds) (1993), *Vulnerability Assessment to Sea Level Rise and Coastal Zone Management.* Proceedings, IPCC/WCC'93 Eastern Hemisphere Preparatory Workshop, Tsukuba, Aug. 1993. Department of Environment, Sport and Territories, Canberra.

Meadows, D.H. Meadows, D.L. and Randers, J. (1992), *Beyond the Limits: Global Collapse or a Sustainable Future,* Earthscan Publications, London.

Meehl, G.A. Washington, W.M. Collins, W.D. et al. (2005), 'How Much More Global Warming and Sea Level Rise?', *Science,* vol. 307, pp. 1769–72.

Meinshausen, M. (2006), 'What Does a 2°C Target Mean for Greenhouse Gas Concentrations? A Brief Analysis Based on Multi-Gas Emission Pathways and Several Climate Sensitivity Uncertainty Estimates', in H.J. Schellnhuber et al. (eds), *Avoiding Dangerous Climate Change,* Cambridge University Press, Cambridge, pp. 265–80.

Mendelsohn, R. (2000), 'Efficient Adaptation to Climate Change', *Climatic Change,* vol. 45, pp. 583–600.

Mendelsohn, R. Morrison, W. Schlesinger, M.E. and Andronova, N.G. (2000), 'Country-Specific Market Impacts of Climate Change', *Climatic Change,* vol. 45, pp. 553–69.

Mendelsohn, R. Nordhaus, W. and Shaw, D. (1996), 'The Impact of Global Warming on Agriculture: A Ricardian Analysis', *American Economic Review,* vol. 84, pp. 753–71.

Messner, S. (1997), 'Endogenized Technological Learning in an Energy Systems Model', *Journal of Evolutionary Economics,* vol. 7, pp. 291–313.

Midun, Z. and Lee, S.C. (1995), 'Implications of a Greenhouse-Induced Sea-level Rise: A National Assessment for Malaysia', *Journal of Coastal Research,* vol. S1, no. 14, pp. 96–115.

Milliman, J.D. and Haq, B.U. (1996), *Sea-level Rise and Coastal Subsidence: Causes Consequences and Strategies,* Kluwer Academic Press, Dordrecht, Netherlands.

Ministry of Science Technology and the Environment Malaysia (2000), *Malaysia: Initial National Communication,* Ministry of Science Technology and the Environment, Kuala Lumpur.

Mishra, H.R. McNeely, J.A. and Thorsell, J.W. (1997), 'Tropical Asia Protects Its Natural Resources', *Forum for Applied Research and Public Policy,* vol. 12, pp. 128–35.

Modelski, G. and Poznanski, K. (1996), 'The Evolutionary Analogy', *International Studies Quarterly,* vol. 4, no. 3.

Morgan, M.G. Kandlikar, M. Risbey, J. and Dowlatabadi, H. (1999), 'Why Conventional Tools for Policy Analysis are Often Inadequate for Problems of Global Change', *Climatic Change,* vol. 41, pp. 271–81.

Mulder, P. Reschke, C.H. and Kemp, R. (1999), *Evolutionary Theorising on Technological Change and Sustainable Development*, Paper prepared for the European Meeting on Applied Evolutionary Economics, 7–9 June.
Munasinghe, M. (1993), *Environmental Economics and Sustainable Development*, Environmental Paper no. 3, World Bank Washington D.C.
Munasinghe, M. (1999), 'Is Environmental Degradation an Inevitable Consequence of Economic Growth: Tunneling Through the Environmental Kuznets Curve', *Ecological Economics*, vol. 29, pp. 89–109.
Munasinghe, M. (2001), *Towards Sustainomics*, Edward Elgar, London.
Munday, S.C.R. (1996), *Current Developments in Economics*, Macmillan Press, London.
Murphy, J.M. Sexton, D.M.H Barnett, D.N. et al. (2004), 'Quantification of Modelling Uncertainties in a Large Ensemble of Climate Change Simulations', *Nature*, vol. 430, pp. 768–72.
Murray, C.J.L and Lopez, D. (eds) (1996), *The Global Burden of Disease: A Comprehensive Assessment of Mortality and Disability from Diseases, Injuries, and Risk Factors in 1990 and Projected to 2020*, Harvard University Press, Cambridge.
NASNAEIM (National Academy of Sciences, National Academy of Engineering and Institute of Medicine) (1992), *Policy Implications of Greenhouse Warming: Mitigation Adaptation, and the Science Base*, National Academy Press, Washington.
NISTEP (National Institute of Science and Technology Policy) (1991), *Analysis of the Structure of Energy Consumption and the Dynamics of Emissions of Atmospheric Species Related to the Global Environmental Change (So_x, No_x & CO_2) in Asia*, NISTEP Report, no. 21, Science and Technology Agency, Tokyo.
Nelson, R.R. and Winter, S. (1982), *An Evolutionary Theory of Economic Change*, Harvard University Press, Cambridge.
Nicholls, R.J. Hoozemans, F.M.J. and Marchand, M. (1999), 'Increasing Flood Risk and Wetland Losses Due to Global Sea-Level Rise: Regional and Global Analyses', *Global Environmental Change*, vol. 9, pp. S69–S87.
Nicholls, R.J. Mimura, N. and Topping, J.C. (1995), 'Climate Change in South and South East Asia; Some Implications for Coastal Areas', *Journal of Global Environmental Engineering*, vol. 1, pp. 137–54.
Nordhaus, W.D. (1967), The Optimal Rate and Direction of Technical Change, in Shell, K. (ed.), *Essays on the Theory of Optimal Economic Growt*h, MIT Press, Cambridge.
Nordhaus, W.D. (1982), 'The Global Commons I: Costs and Climatic Effects, How Fast Should We Graze the Global Commons', *AEA Papers and Proceedings*, vol. 72, no. 2, pp. 242–6.
Nordhaus, W.D. (1989), *The Economics of the Greenhouse Effect*, paper presented to the International Energy Workshop, Laxenburg, 20–22 June.
Nordhaus, W.D. (1991), 'To Slow or Not to Slow: the Economics of the Greenhouse Effect', *Economic Journal*, vol. 101, pp. 920–37.
Nordhaus, W.D. (1993), 'How Much Should We Invest In Preserving Our Current Climate?', in Giersch, H. (ed.), *Economic Progress and Environmental Concerns*, Springer-Verlag, Berlin.

Nordhaus, W.D. (1994a), *Managing the Global Commons: The Economics of Climate Change*, MIT Press, Cambridge.

Nordhaus, W.D., (1994b), 'Expert Opinion on Climate Change', *American Scientist*, vol. 82, pp. 45–51.

Nordhaus, W.D. (1995), 'The Ghosts of Climates Past and the Specters of Climate Change Future', *Energy Policy*, vol. 23, no. 4/5, pp. 269–82.

Nordhaus, W.D. (1997), *Modeling Induced Innovation in Climate-Change Policy*, mimeo, Yale University.

Nordhaus, W.D. and Boyer, J. (2000), *Warming the World: Economic Models of Global Warming*, MIT Press, Cambridge.

Norgaard, R.B. (1994), *Development Betrayed: The End of Progress and a Coevolutionary Revisioning of the Future*, Routledge, London.

OECD (Organisation for Economic Co-operation and Development) (1995), *Global Warming: Economic Dimensions and Policy Responses*, OECD, France.

Office of Environmental Policy and Planning (2000), *Thailand's Initial National Communication under the United Nations Framework Convention on Climate Change*, Ministry of Science, Technology and Environment, Bangkok.

Olmos, S. (2001), *Vulnerability and Adaptation to Climate Change: Concepts Issues Assessment Methods*, Foundation Paper, Climate Change Knowledge Network.

OTA (Office of Technology Assessment) (1991), *Energy in Developing Countries*, United States Congress, Office of Technology Assessment, Washington, D.C.

Page, C. (2001), *Thematic Workshop on Vulnerability and Adaptation Assessment*, National Communications Support Programme Workshop Report, Jakarta, 10–12 May.

Parikh, K.S. (1994), 'Agriculture and Food System Scenarios for the 21st Century', in Ruttan, V.W. (ed.), *Agriculture Environment and Health: Sustainable Development in the 21st Century*, University of Minnesota Press.

Parry, M.L. Blantran, M. de Rozari, Chong, A.L. and Panich, S. (1992), *The Potential Socio-Economic Effects of Climate Change in Southeast Asia*, United Nations Environment Programme, Nairobi.

Parson, E.A. (1995), 'Integrated Assessment and Environmental Policy Making: in Pursuit of Usefulness', *Energy Policy*, vol. 23, no. 4/5, pp. 463–75.

Peck, S.C. and Teisberg, T.J. (1993), 'The Importance of Nonlinearities in Global Warming Damage Costs', in Darmstadter, J. and Toman, M.A. (eds), *Assessing Surprises and Nonlinearities in Greenhouse Warming: Proceedings of an Interdisciplinary Workshop*, Resources for the Future, Washington.

Peck, S.C. and Teisberg, T.J. (1995), 'International CO_2 Emissions Control: An Analysis Using CETA', *Energy Policy*, vol. 23, no. 4/5, pp. 297–308.

Perez, R.T. Feir, R.B. Carandand, E. and Gonzalez, E.B. (1994), 'Potential Impacts of Sea Level Rise on the Coastal Resources of Manila Bay: A Preliminary Vulnerability Assessment', *Water Air and Soil Pollution*, vol. 92, pp. 137–47.

Philippine's Initial National Communication on Climate Change (1999), *The Philippine's Initial National Communication on Climate Change*.

Pielke, R.A. (1998), 'Rethinking the Role of Adaptation in Climate Policy', *Global Environmental Change*, vol. 8, pp. 159–70.
Pierrehumbert, R.T. (2004), 'Warming the World', *Nature*, vol. 432, p. 677.
Plambeck, E.L. Hope, C. and Anderson, J. (1997), 'The PAGE95 Model: Integrating the Science and Economics of Global Warming', *Energy Economics*, vol. 19, pp. 77–101.
Plass, G.N. (1956a), 'Effect of Carbon Dioxide Variations on Climate', *American Journal of Physics*, vol. 24, pp. 376–87.
Plass, G.N. (1956b), 'The Influence of the 15-Micron Carbon Dioxide Band on the Atmospheric Infrared Cooling Rate', *Quarterly Journal of the Royal Meteorological Society*, vol. 82, pp. 310–24.
Plass, G.N. (1956c), 'The Carbon Dioxide Theory of Climatic Change', *Tellus*, vol. 8, pp. 140–54.
Plass, G.N. (1961), 'The Influence of Infrared Absorptive Molecules on the Climate', *Annals of the New York Academy of Science*, vol. 95, pp. 61–71.
PSAC (President's Science Advisory Committee) (1965), *Restoring the Quality of our Environment: Report of the Environmental Pollution Panel*, President's Science Advisory Committee, The White House, Washington, D.C.
Quibria, M.G. (ed.) (1993), *Rural Poverty in Asia: Priority Issues and Policy Options*, Oxford University Press, Hong Kong.
Quibria, M.G. (ed.) (1995), *Critical Issues in Asian Development: Theories, Experiences and Policies*, Oxford University Press, Asian Development Bank, Hong Kong.
Qureshi, A. and Hobbie, D. (eds) (1994), *Climate Change in Asia: Executive Summary*, Asian Development Bank, Manila.
Radzicki, M.J. and Sterman, J.D. (1994), 'Evolutionary Economics and System Dynamics', in England, R.W. (ed.), *Evolutionary Concepts in Contemporary Economics*, The University of Michigan Press, Michigan.
Rebelo, S. (1991), 'Long-Run Policy Analysis and Long-Run Growth', *Journal of Political Economy*, vol. 99, no. 3, pp. 500–21.
Reijnders, J. (ed.) (1997), *Economics and Evolution*, Edward Elgar, Cheltenham.
Reilly, J. (1996), 'Climate Change, Global Agriculture and Regional Vulnerability', in Bazzaz, F. and Sombroek, W. (eds), *Global Climate Change and Agricultural Production: Direct and Indirect Effects of Changing Hydrological, Pedological and Plant Physiological Processes*, FAO and John Wiley & Sons, pp. 237–65.
Reilly, J.M. and Schimmelpfennig, D. (1999), 'Agricultural Impact Assessment Vulnerability and the Scope for Adaptation', *Climatic Change*, vol. 43, pp. 745–88.
Republic of the Philippines (1997), *Philippine Agenda 21: A National Agenda for Sustainable Development*.
Revelle, R. and Seuss, H. (1957), 'Carbon Dioxide Exchange Between the Atmosphere and Ocean and the Question of an Increase of Atmospheric CO_2 During the Past Decade', *Tellus*, vol. 9, no. 18.
Roberts, J.T. and Grimes, P.E. (1997), 'Carbon Intensity and Economic Development 1962–91: A Brief Exploration of the Environmental Kuznets Curve', *World Development*, vol. 25, no. 2, pp. 191–8.

Rogers, P. (1993), 'The Environment in Southeast Asia', *Environmental Science and Technology*, vol. 27, no. 12, p. 2269.

Romer, P.M. (1986), 'Increasing Returns and Long-run Growth', *Journal of Political Economy*, vol. 94, no. 5, pp. 1002–37.

Romer, P.M. (1990), 'Endogenous Technological Change', *Journal of Political Economy*, vol. 98, no. 5, pp. 71–102.

Root, T.L. MacMynowski, D.P. Mastrandrea, M.D. and Schneider, S.H. (2005), 'Human-Modified Temperatures Induce Species Changes: Combined Attribution', *Proceedings of the National Academy of Sciences*, vol. 102, pp. 7465–9.

Rosegrant, M.W. and Ringler, C. (1997), 'World Food Markets into the 21st Century: Environmental and Resource Constraints and Policies', *The Australian Journal of Agriculture and Resource Economics*, vol. 41, no. 3, pp. 401–28.

Rosegrant, M.W. and Svendsen, M. (1993), 'Asian Food Production in the 1990s: Irrigation Investment and Management Policy', *Food Policy*, vol. 18, no. 1, pp. 13–32.

Rosenberg, D. (1999), *Environmental Pollution around the South China Sea: Developing a Regional Response to a Regional Problem*, Working Paper 1999/20, Resource Management in Asia-Pacific Project Seminar Series, Australian National University.

Rosenberg, N.J. (1992), 'Adaptation of Agriculture to Climate Change', *Climatic Change*, vol. 21, pp. 385–405.

Rotmans, J. (1990), *Image: An Integrated Model to Assess the Greenhouse Effect*, Kluwer, Dordrecht.

Rotmans, J. and Dowlatabadi, H. (1996), 'Integrated Assessment of Climate Change: Evaluation of Models and Other Methods', in Rayner, S. and Malone, E. (eds), *Human Choice and Climate Change: An International Social Science Assessment*, Battelle Press, Columbus.

Sanderson, J.R. and Islam, S.M.N. (2000a), 'Climate Change in Asia: Issues and Policy Options', *Natural Resources Forum*, vol. 24, no. 1, pp. 39–48.

Sanderson, J.R. and Islam, S.M.N. (2000b) *Economic Development and Climate Change in South East Asia: Costs, Issues and Policy Options*, Background paper prepared for the 2nd Authors Meeting of the book 'Policy Implications of Global Change for South East Asia', 14–15 January, Chiang Mai.

Sanderson, J. and Islam, S.M.N. (2001), Economic Development and Climate Change in South East Asia: The SEADICE model and its Forecasts for Growth Prospects and Policy Strategies, *International Journal of Global Environmental Issues*, vol. 3, no. 2, 2003, Inderscience.

Schar, C. Vidale, P.L. Luthi, D. et al. (2004), 'The Role of Increasing Temperature Variability in European Summer Heatwaves', *Nature*, vol. 427, pp. 332–6.

Scheffer, M. Brovkin, V. and Cox, P. (2006), 'Positive Feedback between Global Warming and the Atmospheric CO_2 Concentration Inferred from Past Climate Change', *Geophysical Research Letters*, vol. 33.

Schellnhuber, H.J. Cramer, W. Nakicenovic N. et al. (eds) (2006), *Avoiding Dangerous Climate Change*, Cambridge University Press, Cambridge.

Scheraga, J.D. and Grambsch, A.E. (1998), 'Risks Opportunities and Adaptation to Climate Change', *Climate Research*, vol. 10, pp. 85–95.

Schimmelpfennig, D. and Yohe, G. (1999), 'Vulnerability of Crops to Climate Change: A Practical Method of Indexing', in Frisvold, G. and Kuhn, B. *Global Environmental Change and Agriculture: Assessing the Impacts*, Edward Elgar, Cheltenham, pp. 193–217.

Schimmelpfennig, D. Lewandrowski, J. Reilly, J. Tsigas, M. and Parry, I. (1996), *Agricultural Adaptation to Climate Change: Issues of Longrun Sustainability*, United States Department of Agriculture, Economic Research Service, AER-740, Washington D.C.

Schneider, S.H. (1997), 'Integrated Assessment Modeling of Global Climate Change: Transparent Rational Tool for Policy Making or Opaque Screen Hiding Value-laden Assumptions?', *Environmental Modeling and Assessment*, vol. 2, pp. 229–49.

Schneider, S.H. and Thompson, S.L. (1981), 'Atmospheric CO_2 and Climate: Importance of the Transient Response', *Journal of Geophysical Research*, vol. 86, no. C4, pp. 3135–47.

Scholze, M. Wolfgang, K. Arnell, N. and Prentice, C. (2006), 'A Climate-Change Risk Analysis for World Ecosystems', *Proceedings of the National Academy of Sciences*, vol. 103, pp. 13116–20.

Sen, A. (1970), *Growth Economics*, Penguin, Middlesex.

Sen, A. (1995), 'Environmental Evaluation and Social Choice: Contingent Valuation and the Market Analogy', *The Japanese Economic Review*, vol. 46, no. 1, pp. 23–37.

Shafik, N. (1994), 'Economic Development and Environmental Quality: An Econometric Analysis', *Oxford Economic Papers*, vol. 46, pp. 757–73.

Sharma, S. (1994), 'Greenhouse Gases and Energy Policies in the Asia-Pacific', in Sharma, S. (ed.), *Energy, the Environment and the Oil Market*, Institute of Southeast Asian Studies, Singapore.

Sheehan, P. (2000), *Manufacturing and Growth in the Longer Term: An Economic Perspective*, Working Paper no. 17, Centre for Strategic Economic Studies.

Shell, K. (1966), 'Toward a Theory of Inventive Activity and Capital Accumulation', *American Economic Review*, Papers and Proceedings, vol. 56, pp. 62–8.

Sheraga, J. and Grambsch, A. (1998), 'Risks Opportunities and Adaptation to Climate Change', *Climate Research*, vol. 10, pp. 85–95.

Singapore Ministry of the Environment (2000), *Singapore's Initial National Communication: Under the United Nations Framework Convention on Climate Change*, Ministry of the Environment, Singapore.

Smit, B. Burton, J. Klein, R.J.T. and Street, R. (1999), 'The Science of Adaptation: A Framework for Assessment', *Mitigation and Adaptation Strategies for Global Change*, vol. 4, pp. 199–213.

Smit, B. Burton, J. Klein, R.J.T. and Wandel, J. (2000), 'An Anatomy of Adaptation to Climate Change and Variability', *Climatic Change*, vol. 45, pp. 223–51.

Smith, J. and Lenhart, S. (1996), 'Climate Change Adaptation Policy Options', *Climate Research*, vol. 6, pp. 193–201.

Smith, J.B. (1997), 'Setting Priorities for Adapting to Climate Change', *Global Environmental Change*, vol. 7, no. 3, pp. 251–64.

Smith, J.B. Bhatti, N. Menzhulin, G.V. Benioff, R. Campos, M. Jallow, B. Rijsberman, F. Budyko, M.I. and Dixon, R.K. (eds) (1996), *Adapting to Climate Change: An International Perspective*, Springer-Verlag, New York.

Smith, J.B. Huq, S. Lenhart, S. Mata, L.J. Nemesova, I. and Toure, S. (eds) (1996), *Vulnerability and Adaptation to Climate Change: Interim Results from the U.S. Country Studies Program*, Kluwer Academic Publishers, Dordrecht.

Smith, J.B. Ragland, S.E. and Pitts, G.J. (1996), 'A Process for Evaluating Anticipatory Adaptation Measures for Climate Change', *Water Air and Soil Pollution*, vol. 92, pp. 229–38.

Smith, L.C. Sheng, Y. MacDonald, G.M. et al. (2005), 'Disappearing Arctic Lakes', *Science*, vol. 308, p. 1429.

Smithers, J. and Smit, S. (1997), 'Human Adaptation to Climatic Variability and Change', *Global Environmental Change*, vol. 7, no. 2, pp. 129–46.

Smulders, S. (1998), 'Technological Change, Economic Growth and Sustainability', in Bergh, J.C.J.M. and Hofkes, M.W. (eds), *Theory and Implementation of Economic Models for Sustainable Development*, Kluwer Academic Publishers, Dordrecht.

Snooks, G.D. (1998), *Longrun Dynamics: A General Economic and Political Theory*, Macmillan Press, London.

Solow, R. (1957), 'Technical Change and the Aggregate Production Function', *Review of Economics and Statistics*, vol. 39, pp. 312–20.

Stakhiv, E. (1993), *Evaluation of IPCC Adaptation Strategies*, Draft Report, Institute for Water Resources, United States Army Corps of Engineers, Fort Belvoir.

Stern, N. (2006), *Stern Review: The Economics of Climate Change*, available at: http://www.hm-treasury.gov.uk/independent_reviews/stern_review_economics_ climate_change/stern_review_report.cfm

Stern, N. Peters, S. Bakhshi, V. et al. (2006), *Stern Review: The Economics of Climate Change*, HM Treasury, London.

Stratus Consulting (1999), *Compendium of Decision Tools to Evaluate Strategies for Adaptation to Climate Change*, Stratus Consulting.

SCEP (Study of Critical Environmental Problems) (1970), *Man's Impact on the Global Environment: Assessment and Recommendations for Action*, MIT Press, Cambridge.

Sugandy, A. Bey, A. Gunardi, R. Boer, H. Pawitan, S. Wigenasantana, Hidayat, A. and Utomo, P. (2000), *Indonesia: The First National Communication on Climate Change Convention*, State Ministry of Environment Office, Jakarta.

Sugden, R. and Williams, A. (1978), *The Principles of Practical Cost-benefit Analysis*, Oxford University Press, Oxford.

Swan, T.W. (1956), 'Economic Growth and Capital Accumulation', *Economic Record*, vol. 32, pp. 334–61.

Teh, T.S. and Voon, P.K. (1992), 'Impacts of Sea Level Rise in West Johore Malaysia', *Malaysian Journal of Tropical Geography*, vol. 23, no. 2, pp. 93–102.

Titus, J. (1992), 'The Cost of Climate Change to the United States', in Majumdar, S.K. and Rosenfeld, L.S. (eds), *Global Climate Change: Implications Challenges and Mitigation Measures*, Pennsylvania Academy of Science, Pennsylvania.

Tobin, R.J. and White, A.T. (1993), 'Coastal Resources Management and Sustainable Development: A Southeast Asian Development', *International Environmental Affairs*, vol. 5, no. 1, pp. 50–65.

Tol, R. Fankhauser, S. and Smith, J. (1998), 'The Scope for Adaptation to Climate Change: What Can we Learn from the Impact Literature', *Global Environmental Change*, vol. 8, pp. 109–23.

Tol, R.S.J. (1995), 'The Damage Costs of Climate Change Towards More Comprehensive Calculations', *Environmental and Resource Economics*, vol. 5, pp. 353–74.

Tol, R.S.J. (1996), 'The Damage Costs of Climate Change Towards a Dynamic Representation', *Ecological Economics*, vol. 19, pp. 67–90.

Tol, R.S.J. (1999), 'Time Discounting and Optimal Emission Reduction: An Application of FUND', *Climatic Change*, vol. 41, pp. 351–62.

Tomitate, T. (1991), 'Energy and Global Warming Issues: A Japanese View', in ESCAP, *Energy Policy Implications of the Climatic Effects of Fossil Fuel Use in the Asia-Pacific Region*, Bangkok.

Tongyai, C. (1994), 'Impact of Climate Change on Simulated Rice Production in Thailand', in Rosenzweig, C. and Iglesias, A. (eds), *Implications of Climate Change for International Agriculture: Crop Modelling Study*, United States Environment Protection Authority, Washington.

Toth, F.L. (1995), 'Practise and Progress in Integrated Assessments of Climate Change: A Workshop Overview', *Energy Policy*, vol. 23, no. 4/5, pp. 253–67.

Tsigas, M.E. Frisvold, G.B. and Kuhn, B. (1997), 'Global Climate Change and Agriculture', Chapter 11 in Hertel, T.W. (ed.), *Global Trade Analysis: Modeling and Applications*, Cambridge University Press, Cambridge.

Tucker, M. (1995), 'Carbon Dioxide Emissions and Global GDP', *Ecological Economics*, vol. 15, pp. 215–23.

Turner, R.K. Adger, N. and Doktor, P. (1995), 'Assessing the Economic Costs of Sea Level Rise', *Environment and Planning*, vol. 27, pp. 1777–96.

Turton, H. and Hamilton, C. (1999), *Population Growth and Greenhouse Gas Emissions Sources: Trends and Projections in Australia*, Discussion Paper no. 26, The Australia Institute, Canberra.

Tyndall, J. (1863), 'On Radiation Through the Earth's Atmosphere', *Philosophical Magazine* (4th series), vol. 25.

Ulijaszek, S. (1997), 'Human Adaptation and Adaptability', in Ulijaszek, S.J. and Huss-Ashmore, R. (eds), *Human Adaptability: Past Present and Future*, Oxford University Press, Oxford, pp. 7–16.

UNEP Collaborating Centre on Energy and the Environment (1998), *Mitigation and Adaptation Cost Assessment: Concepts Methods and Appropriate Use*, Riso National Laboratory, Denmark.

UNIDO (United Nations Industrial Development Organization) (2001), *Capacity Building for Industrial CDM Projects: An Initial Assessment of Needs in Asian Countries*, Executive Summary of Interim Country Reports, Vienna.

United Nations (1992), *Report of the United Nations Conference on Environment and Development*, Annex I, Rio Declaration on Environment and Development, Rio de Janeiro, 3–14 June 1992, available online at: http://www.un.org/documents/ga/conf151/aconf15126-1annex1.htm.

United Nations, *Demographic Yearbook* (various), United Nations Statistical Division.

Valdes, B. (1999), *Economic Growth: Theory Empirics and Policy*, Edward Elgar, Cheltenham.

von Neumann, J. (1938), translated as 'A Model of General Equilibrium', *Review of Economic Studies*, pp. 1945–6.

von Storch, H.E. Zorita, E. Jones, J.M. et al. (2004), 'Reconstructing Past Climate from Noisy Data', *Science*, vol. 306, pp. 679–82.

Watson, R.T. Zinyowera, M.C. and Moss, R.H. (eds), (1998), *The Regional Impacts of Climate Change: An Assessment of Vulnerability*, Cambridge University Press, Cambridge.

Wexler, L. (1996), *Improving Population Assumptions in Greenhouse Gas Emissions Models*, Working Paper WP-96-99, International Institute for Applied Systems Analysis, Laxenburg.

Whetton, P. (1996), *Development of Climate Scenarios for Southeast Asia*, CSIRO Division of Atmospheric Research, Aspendale.

White, J.C. (ed.) (1996), *Global Climate Change: The Economic Costs of Mitigation and Adaptation*, Elsevier Science Publishing, New York.

Woodward, A. Hales, S. and Weinstein, P. (1998), 'Climate Change and Human Health in the Asia Pacific Region: Who will be Most Vulnerable?', *Climate Research*, vol. 11, pp. 31–8.

World Bank (1997), *World Development Indicators 1997*, The World Bank, Washington.

World Bank (2000), *World Development Indicators*, The World Bank, Washington.

World Bank (2001), *World Development Indicators*, The World Bank, Washington.

World Conservation Monitoring Centre and IUCN (1994), *Protected Areas of the World*, vol. 3, Cambridge, UK.

World Resources Institute (1997), *World Resources 1996–97*, Oxford University Press, New York.

World Resources Institute (1999), *World Resources 1998–99*, Oxford University Press, New York.

World Resources Institute (2001), *World Resources 2000–01*, Oxford University Press, New York.

Yohe, G. (2000), 'Assessing the Role of Adaptation in Evaluating Vulnerability to Climate Change', *Climatic Change*, vol. 46, pp. 371–90.

Yong, S.K.T. (1989), 'Coastal Resource Management in the ASEAN Region: Problems and Directions', in Thai-Eng, C. and Pauly, D. (eds), *Coastal Area Management in Southeast Asia: Policies Management Strategies and Case Studies*, International Center for Living Aquatic Resources Management, Manila.

Index